Lecture Notes in Chemistry 74

Springer-Verlag Berlin Heidelberg GmbH

M. Defranceschi   C. Le Bris

Mathematical Models and Methods for Ab Initio Quantum Chemistry

 Springer

Authors

Dr. Mireille Defranceschi
CEA-Saclay, Bât. 125
DPE/SPCP/LEPCA,
91191 Gif-sur-Yvette Cedex, France

E-mail: defrancesc@carnac.cea.fr

Prof. Claude Le Bris
C.E.R.M.I.C.S.
Ecole Nationale des Ponts et Chaussées
6&8 Avenue Blaise Pascal
Cité Descartes, Champs sur Marne
77455 Marne La Vallée Cedex 2, France

E-mail: lebris@cermics.enpc.fr

Cataloging-in-Publication Data applied for

Die Deutsche Bibliothek - CIP-Einheitsaufnahme

Mathematical models and methods for ab initio quantum chemistry /
Mireille Defranceschi ; Claude LeBris. - Berlin ; Heidelberg ; New
York ; Barcelona ; Hong Kong ; London ; Milan ; Paris ; Singapore ;
Tokyo : Springer, 2000
 (Lecture notes in chemistry ; 74)
 ISBN 978-3-540-67631-7   ISBN 978-3-642-57237-1 (eBook)
 DOI 10.1007/978-3-642-57237-1
ISSN 0342-4901
ISBN 978-3-540-67631-7

© Springer-Verlag Berlin Heidelberg 2000
Originally published by Springer-Verlag Berlin Heidelberg New York in 2000
Softcover reprint of the hardcover 1st edition 2000

Typesetting: Camera ready by author
Printed on acid-free paper   SPIN: 10771679   51/3142 - 543210

Contents

Foreword

On the occasion of the fourth International Conference on Industrial and Applied Mathematics[1], we decided to organize a sequence of 4 minisymposia devoted to the mathematical aspects and the numerical aspects of Quantum Chemistry. Our goal was to bring together scientists from different communities, namely mathematicians, experts at numerical analysis and computer science, chemists, just to see whether this heterogeneous set of lecturers can produce a rather homogeneous presentation of the domain to an uninitiated audience.

To the best of our knowledgde, nothing of this kind had never been tempted so far. It seemed to us that it was the good time for doing it, both ·because the interest of applied mathematicians into the world of computational chemistry has exponentially increased in the past few years, and because the community of chemists feels more and more concerned with the numerical issues.

Indeed, in the early years of Quantum Chemistry, the pioneers (Coulson, Mac Weeny, just to quote two of them) used to solve fundamental equations modelling toy systems which could be simply numerically handled in view of their very limited size. The true difficulty arose with the need to model larger systems while possibly taking into account their interaction with their environment. Hand calculations were no longer possible, and computing science came into the picture. Today, the challenge is twofold: improving the formalism (both from a physical viewpoint and from the mathematical viewpoint), and speeding up rigorously founded numerical methods. From this originates the revival of interest in the communities of chemists and applied mathematicians.

From what we heard from the audience of our series of minisymposia, the result of our enterprise was beyond our hope. So the idea came out to translate that in a written manner. We therefore suggested that each of the nineteen lecturers should give a written account on his view on the subject. The topic of the written contribution need not be exactly the same as that of the talk, but it has to be related with the interplay between Mathematics, Numerical Analysis and Quantum Chemistry. The volume the two of us planned to edit was not a proceedings volume. It was rather an outgrowth of the series of talks, and an attempt to bring together this heterogeneous population this time in the written mode. The only constraint we imposed to the contributors was the following rule of the game: the mathematicians had to write in a language understandable by chemists, and *vice-versa*. Thirteen lecturers out of nineteen played the game until the end. The result is in the reader's

---

[1]ICIAM99, held in Edinburgh, Scotland, July 5-9th, 1999,

hands. It consists of eleven chapters devoted to various aspects of Computational Chemistry.

This volume is divided into three parts. The first part deals with topics of general interest, the second one is devoted to the modelling of the condensed phases, the third one focuses on the relativistic aspects.

The book opens with a contribution by Brian Sutcliffe on questions of symmetry in Quantum Mechanics. We find it symbolic that the first chapter of such a volume is written by such an eminent chemist, who is known to have had a constant interest into the mathematical aspects.

The first part continues with a chapter by Eric Cancès on the numerical analysis of SCF algorithms for HF calculations. Eric Cancès is one of the representatives of this new generation of applied mathematicians who are very much involved in Computational Chemistry. He presents a rigorous analysis of the existing SCF algorithms and introduces new ones.

Chapter 3 is written together by a chemist and a mathematician, Michel Caffarel and Roland Assaraf. It is devoted to Quantum Monte-Carlo methods. Their contribution, that we consider in some sense as an instance of what should be done in order to enhance the links between the two communities, has been deliberately written on a pedagogical tone. They are to be thanked for that (unfortunately unusual but so much useful) intention.

The last chapter of the first part is due to Gabriel Turinici, who introduces the reader to the very important issue of exact control of quantum systems.

As announced above, the second part of this volume deals with the modelling of the condensed phases. In Chapters 5 to 8, the crystalline solid phase is of concern. Chapter 9 is devoted to the liquid phase.

Chapters 5 and 6 rather stand on the theoretical side. The former, due to Isabelle Catto, Pierre-Louis Lions and one of us (CLB), reports on a series of works devoted to the rigorous derivation of the models of the solid phase, a topic not so often addressed in the physical literature. The latter, written by Olivier Bokanowski, Benoît Grébert and Norbert Mauser, follows the same vein. It discusses the rigorous foundations of some well accepted approximations of Quantum Chemistry, such as the Xα method.

In Chapter 7, Xavier Blanc, a young mathematician, presents his personal view on the models in use for the simulation of the crystalline phase. It may be interesting for a chemist expert at this topic to see the presentation of it by someone of the "outer world".

Chapter 8, from Vanina Louis-Achille and one of us (MD), is the logical sequel of Chapter 7; it develops the practical numerical aspects encountered in solid quantum chemistry calculations which are hidden in usual scientific papers.

Chapter 9 by Benedetta Mennucci is an overview of liquid state methods of calculations. It provides the theoretical framework for the various methods currently used in Quantum Chemistry codes.

The third part, devoted to the relativistic models, features two contributions.

Chapter 10 is due to Jean Dolbeault, Maria Esteban and Eric Séré. It reports on the works of the authors that give a sound mathematical ground to the Dirac-Fock model. By their contribution to the field, they bring the mathematical understanding of the Dirac-Fock model to the level to that of the Hartree-Fock model.

Trond Saue and H. J. Aa. Jensen introduce in Chapter 11 the various concepts of quaternion symmetry appearing in the Dirac Equations as they are implemented in the 4-component relativistic molecular calculations (DIRAC code).

Let us end this foreword by emphasizing we strongly believe that the two communities of applied mathematicians and chemists can take benefit of an interaction. At least we hope that our endeavour towards the challenge to bring together applied mathematicians and chemists will stimulate other ones. As we are tenacious, we are currently working on another project in the same vein.

Needless to say, we encourage every interested reader to contact us.

M. D. & C. L. B.
Paris, April 2000.

Part I

General topics

Chapter 1

Is a molecule in chemistry explicable as a broken symmetry in quantum mechanics?

Brian Sutcliffe
Theoretical Division 1,
Institute for Molecular Science,
Myodaiji, Okazaki 444-8585, Japan,
and present address
Department de Chimie Physique Moléculaire,
Université Libre de Bruxelles,
B-1050 Bruxelles, Belgium.
`bsutclif@ulb.ac.be`

Abstract : It is argued that incorporating the molecular geometry into the transformation specifications for deriving the standard (Eckart) Hamiltonian used to describe molecular spectra, cannot generally be accomplished without breaking the permutational symmetry requirements on identical nuclei.

1   Introduction

The idea that the proper way to treat molecules in quantum mechanics is to treat their nuclear motions as essentially classical with the electrons treated quantum mechanically, dates from the very earliest days of the subject. The genesis of the idea is usually attributed to Born and Oppenheimer [1], but it is an idea that was in the air at the time, for the earliest papers in which the idea is used, predate the publication of their paper. The physical picture that informs the attempted separation is one well known and widely used even in classical mechanics, namely division of the problem into a set of rapidly moving particles, here electrons, and a much more slowly moving set, here the nuclei. Experience is that it is wise to try and separate such incommensurate motions both to calculate efficiently and to get a useful physical picture.

The object of the separation in the molecular case is to get an electronic motion problem in which the nuclear positions can be treated as parameters and whose solutions can be used to solve the nuclear motion problem. The insights arising from classical chemistry seem to predicate that, for lowish energies, the nuclear motion function should be strongly peaked at a nuclear geometry that corresponds to the traditional molecular geometry. A function of this kind would allow a good account of the electronic structure of a molecule to be given in terms of a single choice for the nuclear geometry about which a molecule could be considered as performing small vibrations and undergoing essentially rigid rotations. Carl Eckart [2] was among the first to discuss how this picture might be supported. He did so in a context that assumed the separation of electronic and nuclear motions. In his approach, the electrons are regarded simply as providing a potential. This potential is invariant under all uniform translations and rigid rotations of the nuclei that form the molecule. It is usually referred to as a potential energy surface and the nuclei are said to move on this surface. Eckart actually treats the nuclear motions by classical rather than quantum mechanics but it is his approach and developments from it that have dominated the interpretation of molecular spectra since 1936. Because the approach is essentially classical, it is perfectly proper to treat identical particles as distinguishable, by virtue of their positions on the potential surface. This identification allows the assignment of a molecule to a point group, that of its equilibrium geometry, and such assignments have proved extremely fruitful in interpreting molecular vibration-rotation spectra.

In what follows we shall not concern ourselves with the details of the separation of electronic from nuclear motion but rather with the consequences of trying to treat the nuclear motion quantum mechanically rather than classically and to allow for the permutational symmetry of identical nuclei and for the overall translation and rotation symmetries.

2   The equations of motion for the molecule

Schrödinger's Hamiltonian describing the molecule as a system of N charged particles in a coordinate frame fixed in the laboratory is

$$\hat{H}(\mathbf{x}) = -\frac{\hbar^2}{2} \sum_{i=1}^{N} \frac{1}{m_i} \nabla^2(\mathbf{x}_i) + \frac{e^2}{8\pi\epsilon_o} \sum_{i,j=1}^{N} {}' \frac{Z_i Z_j}{x_{ij}} \tag{1}$$

where the separation between particles is defined by

$$x_{ij}^2 = \sum_{\alpha} (x_{\alpha j} - x_{\alpha i})^2 \tag{2}$$

It is convenient to regard $\mathbf{x}_i$ as a column matrix of three cartesian components $x_{\alpha i}$, $\alpha = x, y, z$ and to regard $\mathbf{x}_i$ collectively as the 3 by N matrix $\mathbf{x}$. Each of the particles has mass m_i and charge $Z_i e$. The charge-numbers Z_i are positive for a nucleus and minus one for an electron. In a neutral system the charge-numbers sum to zero.

To distinguish between electrons and nuclei, the variables are split up into two sets, one set consisting of L variables, $\mathbf{x}_i^e$, describing the electrons and the other set of H variables, $\mathbf{x}_i^n$, describing the nuclei and $N = L + H$. When it is necessary to emphasise this split, (1) will be denoted $\hat{H}(\mathbf{x}^n, \mathbf{x}^e)$.

If the full problem has eigenstates that are square-integrable so that

$$\hat{H}(\mathbf{x}^n, \mathbf{x}^e)\psi(\mathbf{x}^n, \mathbf{x}^e) = E_{BS}\psi(\mathbf{x}^n, \mathbf{x}^e) \tag{3}$$

then an eigenstate can be written approximately as a product of the required form

$$\psi(\mathbf{x}^n, \mathbf{x}^e) = \Phi(\mathbf{x}^n)\phi(\mathbf{x}^n, \mathbf{x}^e)$$

The function, $\Phi(\mathbf{x}^n)$ (chosen to be non-trivial) is determined as a solution of an effective nuclear motion equation, and traditionally, a good guess for the electronic wave function is supposed to be provided by a solution of the clamped nuclei electronic Hamiltonian

$$\hat{H}^{cn}(\mathbf{a}, \mathbf{x}^e) = -\frac{\hbar^2}{2m} \sum_{i=1}^{L} \nabla^2(\mathbf{x}_i^e) - \frac{e^2}{4\pi\epsilon_0} \sum_{i=1}^{H} \sum_{j=1}^{L} \frac{Z_i}{|\mathbf{x}_j^e - \mathbf{a}_i|} + \frac{e^2}{8\pi\epsilon_0} \sum_{i,j=1}^{N} {}' \frac{1}{|\mathbf{x}_i^e - \mathbf{x}_j^e|} \tag{4}$$

This Hamiltonian is obtained from the original one (1) by assigning the values $\mathbf{a}_i$ to the nuclear variables $\mathbf{x}_i^n$, hence the designation *clamped nuclei* for this form. Within the electronic problem each nuclear position $\mathbf{a}_i$ is treated as a parameter. For solution of the entire problem, the electronic wave function must be available for all values of these parameters. The energy obtained from the solution of this problem depends on the nuclear parameters and is commonly called the electronic energy. It is usual to think of the potential energy surface, used in the Eckart approach, as formed by adding the electronic and the classical nuclear repulsion energy.

But it is not actually possible to use this approach directly to approximate solutions, because the Hamiltonian (1) is invariant under uniform translations in the frame fixed in the laboratory. This means that the centre of molecular mass moves through space like a free particle and the states of a free particle are not quantised and eigenfunctions are not square integrable. Before we consider this further let us note another two symmetries of the equation. The molecular Hamiltonian is invariant under all orthogonal transformations (rotation-reflections) of the particle variables in the frame fixed in the laboratory and is also invariant under the permutation of the variable sets of all identical particles.

The way in which translational motion can be removed from the problem is well understood from classical mechanics. But it involves an essentially arbitrary choice of translationally invariant coordinates. If we want to maintain a distinction between electronic and nuclear variables, it seems sensible to define L translationally invariant electronic coordinates such that they transform into one another in the standard way under electronic permutations and are unaffected by nuclear permutations. The rotational motion should then be described in a translationally invariant way, entirely in terms of the original nuclear coordinates. This process consists essentially of the choice of an orthogonal matrix $\mathbf{C}$ fixing a frame in the body of the molecule along with the choice of $3H - 6$ internal coordinates q_k which describe the changes in nuclear geometry and are invariant under any orthogonal transformation of the original coordinate set. The matrix $\mathbf{C}$ can be parameterised by three Euler angles which may be used to define the angular momentum of the system while the sign of $\det|\mathbf{C}|$ specifies the handedness of the chosen frame. The coordinate transformation from the frame fixed in the laboratory to one fixed in the molecule is clearly a non-linear one and so must fail for some coordinate specification where the jacobian vanishes. So, for example, there will always be a molecular geometry at which, for any particular choice of embedding rule, it is not possible to define three Euler angles and here the transformation will fail. But there will always be a domain of validity and we shall confine attention to that domain. Within that domain the wave function for the problem has the form

$$T(\mathbf{X}_T) \sum_{k=-J}^{+J} \Phi_k^J(\mathbf{q}, \mathbf{z}) |JMk> \tag{5}$$

where $\mathbf{X}_T$ denotes the centre of mass coordinate. and $\mathbf{q}$ and $\mathbf{z}$ the internal nuclear and electronic coordinates respectively while $|JMk>$ is an angular momentum eigenfunction depending on the three Euler angles. The translational motion can be completely separated off but the rotational motion can be separated only approximately. If the separation between electronic and nuclear motion is valid then the internal motion function can be written in good approximation as

$$\Phi_{kp}^J(\mathbf{q})\psi_p(\mathbf{q}, \mathbf{z}) \tag{6}$$

for any electronic state p and it is perfectly reasonable now to regard the electronic wave function as a solution of the clamped nuclei problem, a particular choice of the $\mathbf{a}_i$ yielding a particular choice of the q_k.

A permutation of the variables describing a set of identical nuclei will naturally leave the centre of mass coordinate unchanged and therefore will not affect the removal of translational motion. It will also leave the centre of nuclear mass coordinate unchanged. But it is perfectly possible that such a permutation will induce changes in the orientation variables and in the internal coordinates and it might well be the case that it actually mixes the orientation variables and the internal coordinates.

3   The Eckart choice of a frame fixed in the body

In the Eckart approach to a fixing a frame in a molecule, the electrons are considered simply as providing a potential for nuclear motion and so in discussing this embedding we shall ignore them. Implicitly however, the electronic coordinates are treated relative to the centre of nuclear mass so that the translationally invariant electronic coordinates are just like the original electronic coordinates in that there are just L of them, they are invariant under any permutations of the nuclei and transform in the standard way under electronic permutations. It is perhaps sufficiently clear from this that they cause no deep complications.

The frame $\mathbf{C}$ fixed in the body is defined entirely in terms of the original nuclear coordinates and is such as to define a redundant set of cartesian coordinates $\mathbf{z}_i^n$ according to

$$\mathbf{x}_i^n - \mathbf{X} = \mathbf{C}\mathbf{z}_i^n$$

where $\mathbf{X}$ is the centre of nuclear mass coordinate. Thus

$$\sum_{i=1}^{H} m_i \mathbf{z}_i^n = 0 \tag{7}$$

and the matrix $\mathbf{C}$ is chosen as to define a set of cartesians in the frame fixed in the body such that the reference structure is specified by

$$\mathbf{z}_i^n = \mathbf{a}_i$$

where the $\mathbf{a}_i$ are constant matrices and, by definition,

$$\sum_{i=1}^{H} m_i \mathbf{a}_i = 0$$

The reference structure is, of course, usually chosen to be the classical molecular geometry. It is assumed chosen to reflect the equilibrium geometry of the molecule as it would be at the minimum of the potential. However for the purposes of carrying through the derivations, there is no need to specify it more closely than as a reference configuration.

Defining the displacement coordinates as

$$\boldsymbol{\rho}_i = \mathbf{z}_i^n - \mathbf{a}_i$$

the specification of the matrix $\mathbf{C}$ is completed by requiring that

$$\sum_{i=1}^{H} m_i \vec{a}_i \times \vec{\rho}_i = \vec{0} \tag{8}$$

provided that the $\mathbf{a}_i$ do not define a line. If they do, then they provide only one rather than the required three constraints and the Euler angles of the Eckart frame cannot then be defined. Even when they do not define a line, it is possible that the displacement vectors might, and if they do, the definition of the Eckart frame again will fail. But we assume that this happens only for very large displacements from the reference molecular geometry and thus in a region where the vibrational wave function vanishes.

These constraint conditions are on the components of the mass weighted sum over all the vectors, of the vector products of the reference geometry vectors with the displacement vectors. In classical mechanics the vanishing of these components would be interpreted as the system having no internal angular momentum at the reference geometry. The two conditions (7) and (8) defining the embedded frame are often called the Eckart conditions. The standard account of the Eckart formulation in quantum mechanical form is provided by Watson [3] and, for comparison, here $\mathbf{z}_i^n$ is used for Watson's $\mathbf{r}_i$ and $\mathbf{a}_i$ for his $\mathbf{r}_i^0$.

The $\mathbf{z}_i^n$ are completely expressible in terms of a set of $3H - 6$ internal coordinates q_k. The internal coordinates are expressed in terms of displacements from the reference geometry by

$$q_k = \sum_{i=1}^{H} \sum_{\alpha} b_{\alpha i k} \rho_{\alpha i} \equiv \sum_{i=1}^{H} \mathbf{b}_{ik}^T \boldsymbol{\rho}_i, \quad k = 1, 2, \cdots 3H - 6 \tag{9}$$

where the elements $b_{\alpha i k}$ are simply constants which may be regarded as components of a column matrix $\mathbf{b}_{ik}$. The range of these coordinates is $(-\infty, \infty)$.

Internal coordinates must be linearly independent and so the $\mathbf{b}_{ik}$ must be linearly independent and in order for an inverse transformation to exist between the internal coordinates and the coordinates in the frame fixed in the laboratory, it is [4] required that

$$\sum_{i=1}^{H} \mathbf{b}_{ik} = \mathbf{0} \qquad \sum_{i=1}^{H} \vec{a}_i \times \vec{b}_{ik} = \vec{0}$$

Using the two Eckart conditions given above it follows that

$$\mathbf{C} = \mathbf{B}(\mathbf{B}^T\mathbf{B})^{-\frac{1}{2}} \tag{10}$$

where

$$\mathbf{B} = \sum_{i=1}^{H} m_i(\mathbf{x}_i^n - \mathbf{X})\mathbf{a}_i^T \tag{11}$$

The 3 by 3 matrix $\mathbf{B}^T\mathbf{B}$ is symmetric and therefore diagonalisable, so functions of it may be properly defined in terms of its eigenvalues. There are in principle, eight (2^3) possible distinct square root matrices, all such that their square yields the original matrix product. Consistency (see [5]) requires, however, that the positive square roots of each of the eigenvalues be chosen. The eigenvalues cannot be negative but one or more might be zero and if this is so then the Eckart frame cannot be properly defined This is a matter to which we shall return later.

If we denote the 3 by H matrix of the $\mathbf{x}_i^n - \mathbf{X}$ as $\mathbf{w}$ and the similar collection of all the $\mathbf{z}_i^n$ and thus the $\mathbf{a}_i$ as $\mathbf{z}^n$ and $\mathbf{a}$ then, from (11), $\mathbf{B}$ may be written as

$$\mathbf{B} = \mathbf{wma}^T = \mathbf{Cz}^n\mathbf{ma}^T$$

where $\mathbf{m}$ is an H by H diagonal matrix with the nuclear masses along the diagonal. The second Eckart condition (8) can be manipulated to show that it is equivalent to the requirement that the matrix

$$\mathbf{A} = \sum_{i=1}^{H} m_i \mathbf{z}_i^n \mathbf{a}_i^T \equiv \mathbf{z}^n \mathbf{ma}^T \tag{12}$$

is symmetric. Thus

$$\mathbf{B}^T\mathbf{B} \equiv \mathbf{A}^T\mathbf{A} = \mathbf{A}^2$$

and so from (10)

$$\mathbf{C} = \mathbf{B}(\mathbf{A}^2)^{-\frac{1}{2}}$$

Whatever the precise specification made of the inverse square root, if $\mathbf{A}$ is singular then $\mathbf{C}$ is undefined and the Eckart specification fails. Thus, as mentioned above, if the $\mathbf{a}_i$ together specify a line, for example if all the a_{yi} and a_{zi} vanish then $\mathbf{A}$ is clearly singular. But there is a special case. If the $\mathbf{a}_i$ together specify a planar figure, for example if all the a_{zi} vanish, then $\mathbf{A}$ is again singular but in this case the Eckart conditions may be satisfied by requiring that the z_{zi}^n vanish too. The last row and column of $\mathbf{A}$ are null and there is a 2 by 2 non-vanishing block and, provided that this block is non-singular, the problem remains well defined. As a matter of fact there is no need in practice to treat this special case explicitly. The second Eckart condition implicitly orients the third axis to be perpendicular to the defined plane and, as long as the two by two block of $\mathbf{A}$ is non-singular, the planarity constraints are subsumed.

It is possible to write the cartesians expressed in the Eckart frame directly in terms of the coordinates of the frame fixed in the laboratory as:

$$\mathbf{z}^n = \mathbf{C}^T\mathbf{w} = (\mathbf{amw}^T\mathbf{wma}^T)^{-\frac{1}{2}}\mathbf{amw}^T\mathbf{w}$$

and since the ij-th element of $\mathbf{w}^T\mathbf{w}$ is a scalar product, $(\mathbf{x}_i^n - \mathbf{X})^T(\mathbf{x}_j^n - \mathbf{X})$, the internal cartesians are clearly invariant under any orthogonal transformation, including inversion, of the original nuclear coordinates.

3.1 The effects of nuclear permutations on the Eckart coordinates

If we represent a particular permutation of identical nuclei by $\mathbf{P}$, where this is a standard orthogonal permutation matrix, then we can express any permutation of nuclei by

$$\mathbf{x}^n \rightarrow \mathbf{x}^n\mathbf{P}$$

and we see that this permutation induces in $\mathbf{B}$ the change

$$\mathbf{B} \to \mathbf{w}\mathbf{P}^T\mathbf{m}\mathbf{a}^T = \mathbf{w}\mathbf{m}(\mathbf{a}\mathbf{P})^T = \mathbf{C}\mathbf{z}^n\mathbf{m}(\mathbf{a}\mathbf{P})^T = \mathbf{C}\mathbf{A}^P$$

because the permutation is of identical nuclei, and so its representative matrix must commute with the mass matrix. Thus the permutation seems to act as if it changed the reference vectors but these are, of course, quantities fixed at the time of problem specification and hence unchanging parameters once a choice is made. From (10) it follows that under this permutation

$$\mathbf{C} \to \mathbf{C}\mathbf{A}^P(\mathbf{A}^{P^T}\mathbf{A}^P)^{-\frac{1}{2}} \equiv \mathbf{C}\mathbf{U}$$

If the matrix $\mathbf{A}^P$ is singular then $\mathbf{U}$ will not exist, the change in $\mathbf{C}$ will be undefined, and hence the Eckart frame embedding will fail. So no such permutation can be realised in the Eckart formulation and a fundamental symmetry of of the problem is broken.

If the permutation is such is that $\mathbf{A}^P$ is non-singular and symmetric then it can be diagonalised by an orthogonal matrix $\mathbf{V}$ with eigenvector columns $\mathbf{v}_i$ and eigenvalues a_i^P. In dealing with the square root, consistency again requires the choice of the positive value and $\mathbf{U}$ can therefore be written as

$$\mathbf{U} = \sum_{i=1}^{3} sgn(a_i^P)\mathbf{v}_i\mathbf{v}_i^T$$

If all the a_i^P are positive then $\mathbf{U}$ is just the unit matrix and if all negative then $\mathbf{U}$ is the negative unit matrix and represents an inversion. The form that it actually takes depends on the precise nature of the permutation.

If $\mathbf{U}$ is a constant matrix, separation between angular and internal coordinates is essentially preserved. The most that can happen is that the internal coordinates go into linear combinations of themselves and the angular momentum components go into linear combinations of themselves. A special case of such a constant matrix is if the permutation is such that

$$\mathbf{a}\mathbf{P} \equiv \mathbf{a}^P = \mathbf{S}\mathbf{a}, \quad \mathbf{S}^T\mathbf{S} = \mathbf{E}_3 \tag{13}$$

then

$$\mathbf{C} \to \mathbf{C}\mathbf{S}^T, \quad \mathbf{z}^n \to \mathbf{S}\mathbf{z}^n\mathbf{P}^T \tag{14}$$

The matrix $\mathbf{S}$ could arise as the orthogonal representation matrix of a point group operation on the molecular framework, for the point group of the molecular framework is isomorphic with a subgroup of the full permutation group of the molecule. The subgroup might be the full group, but that is only rarely the case. It is also the case a relation like (13) might be possible for a permutation or set of permutations that are unconnected with any point group operations. When the permutations can be realised by proper or improper rotation matrices, they are often called *perrotations*, a name introduced by Gilles and Philippot [6]. These matters and many other related ones, are discussed in more detail in Chapters 2 and 3 of Ezra's monograph [5].

In general the permutation will yield an unsymmetric $\mathbf{A}^P$ but provided it is non-singular then $\mathbf{U}$ will be well defined and orthogonal.

An example

As an example we shall choose ethene, which also illustrates the planar special case too:

$$
\begin{array}{ccc}
H_1 & & H_4 \\
\backslash & & / \\
& C_1 = C_2 & \\
/ & & \backslash \\
H_2 & & H_3
\end{array}
$$

with the following choice of a_i

$$
a_1 = \begin{pmatrix} -a \\ b \\ 0 \end{pmatrix} \quad
a_2 = \begin{pmatrix} -a \\ -b \\ 0 \end{pmatrix} \quad
a_3 = \begin{pmatrix} a \\ -b \\ 0 \end{pmatrix} \quad
a_4 = \begin{pmatrix} a \\ b \\ 0 \end{pmatrix}
$$

$$
a_5 = \begin{pmatrix} -d \\ 0 \\ 0 \end{pmatrix} \quad
a_6 = \begin{pmatrix} d \\ 0 \\ 0 \end{pmatrix}
$$

Where $a_1 \ldots a_4$ represent the reference positions of the protons, each with mass m_p and a_5 and a_6 the reference positions of the carbons, each with mass m_c. The system is specified to lie in the $x - y$ plane and we shall be concerned only with the 2 by 2 non-vanishing sub matrix of $\mathbf{A}$. The point group of the reference figure is D_{2h}, using the Schönfliess notation which is standard in molecular spectroscopy. The full permutation group is $S_4 \times S_2$ assuming the nuclei to be identical isotopes of hydrogen and of carbon respectively. The point group contains 8 operations while the permutation group contains 48 operations.

It is convenient to rewrite $\mathbf{A}$ from (12) as

$$
m_p z_p^n a_p^T + m_c z_c^n a_c^T \equiv \mathbf{A}_p + \mathbf{A}_c
$$

where the split has been made into the proton and the carbon parts of the problem and permutations will occur only within each part. The carbon part of the problem is always of the form

$$
\begin{pmatrix} \pm m_c(z_{x6}^n - z_{x5}^n)d & 0 \\ 0 & 0 \end{pmatrix} = \pm \mathbf{A}_c
$$

where the plus sign represents the identity permutation and the minus sign the transposition of the carbons. If $\mathbf{A}_p$ is non-singular then

$$
\mathbf{A}_p \pm \mathbf{A}_c = \mathbf{A}_p(\mathbf{E}_2 \pm \mathbf{A}_p^{-1}\mathbf{A}_c)
$$

and even though $\mathbf{A}_c$ is singular there is no reason to expect the second term on the right to be singular, except perhaps for particular values of the coordinates and we can assume that such regions of coordinate space may be treated as unvisited for our present purposes. It is possible that even if $\mathbf{A}_p$ is singular $(\mathbf{A}_p \pm \mathbf{A}_c)$ might not be, but this again will happen only in rather special regions. What has been said in relation to $\mathbf{A}_p$ here, will hold equally for any $\mathbf{A}_p^P$. We would anticipate therefore that that any permutation of the protons that yields a non-singular $\mathbf{A}_p^P$ would yield a non-singular $\mathbf{A}^P$ when taken with either of the possibilities for $\mathbf{A}_c$. If the permuted form is symmetric, then $\mathbf{A}^P$ will be symmetric, though there is no reason to expect this to be generally the case. With these matters in mind we shall concentrate on the proton permutations in what follows.

Looking at the effect of the point group operations on the framework vectors it is easy to see that

$$C_{2x} \equiv \sigma_{zz} \equiv (12)(34)(5)(6), \quad C_{2y} \equiv \sigma_{yz} \equiv (14)(23)(56), \quad C_{2z} \equiv i \equiv (13)(24)(56)$$

while σ_{xy} is equivalent to the identity operation in the point group and the identity permutation. Thus the effect of these permutations on $\mathbf{A}_p$ is

$$\mathbf{A}_p \overset{C_{2x}}{\to} \mathbf{A}_p \begin{pmatrix} 1 & 0 \\ 0 & -1 \end{pmatrix} \quad \mathbf{A}_p \overset{C_{2y}}{\to} \mathbf{A}_p \begin{pmatrix} -1 & 0 \\ 0 & 1 \end{pmatrix}$$

$$\mathbf{A}_p \overset{C_{2z}}{\to} \mathbf{A}_p \begin{pmatrix} -1 & 0 \\ 0 & -1 \end{pmatrix}$$

Adjoining the relevant transformed $\mathbf{A}_c$ to each of these matrices allows the constant matrix to be separated as $\mathbf{S}$ and we obtain precisely the results for $\mathbf{U}$ anticipated from (13) above.

The proton permutations above are the complete set of double transpositions for the protons, and we would expect each of them, even when joined with the alternative carbon permutations to those of the point group, to yield non-singular $\mathbf{A}^P$. Hence the definition of the Eckart frame would be maintained even for such non-point group permutations. However one would not in general be able to factor out a common constant matrix for such permutations as one can for perrotations and one would expect the relevant $\mathbf{U}$ to be functions of the internal coordinates.

It is difficult to analyse the problem further at this level of generality. However if we choose the reference values for the $\mathbf{z}_i^n$ we can go a little further and if, with this special choice, the Eckart frame definition fails, then the frame certainly cannot be defined for small displacements, which form the most important region for the Eckart approach. For present purposes, therefore, we shall regard $\mathbf{A}$ as if composed $\mathbf{ama}^T$ and the transformation as

$$\mathbf{ama}^T \to \mathbf{am}(\mathbf{aP})^T$$

With these restrictions

$$\mathbf{A}_p = \begin{pmatrix} 4m_p a^2 & 0 \\ 0 & 4m_p b^2 \end{pmatrix} \text{ and } \mathbf{A}_c = \begin{pmatrix} 2m_c d^2 & 0 \\ 0 & 0 \end{pmatrix}$$

Looking at the six proton transpositions, four yield singular $\mathbf{A}_p^P$ but two, (13) and (24), do not. The permutation (12) gives

$$\mathbf{a}_p \mathbf{m}_p \mathbf{a}_p^T \to \mathbf{a}_p \mathbf{m}_p \mathbf{a}_p^{P_{12}T} = \begin{pmatrix} 4m_p a^2 & 0 \\ 0 & 0 \end{pmatrix}$$

and the matrix for (34) is the same. In neither case can the singularity be removed with the aid of $\mathbf{A}_c$. The permutation (14) gives

$$\mathbf{a}_p \mathbf{m}_p \mathbf{a}_p^T \rightarrow \mathbf{a}_p \mathbf{m}_p \mathbf{a}_p^{P_{14}T} = \begin{pmatrix} 0 & 0 \\ 0 & 4m_p b^2 \end{pmatrix}$$

and the matrix for (23) is the same. In both cases the singularity can be removed with the aid of $\mathbf{A}_c$. The permutation (24) yields

$$\mathbf{a}_p \mathbf{m}_p \mathbf{a}_p^T \rightarrow \mathbf{a}_p \mathbf{m}_p \mathbf{a}_p^{P_{24}T} = \begin{pmatrix} 0 & -4m_p ab \\ -4m_p ab & 0 \end{pmatrix}$$

and (13) is the same but with positive off-diagonal elements. Both matrices are non-singular and the Eckart frame can certainly be well defined in the case of either of these permutations associated with either of the carbon permutations.

The eight permutations involving just three protons all lead to singular $\mathbf{A}_p^P$ of the form

$$\begin{pmatrix} 0 & \pm 4m_p ab \\ 0 & 0 \end{pmatrix} \text{ or } \begin{pmatrix} 0 & 0 \\ \pm 4m_p ab & 0 \end{pmatrix}$$

so for none of these permutations can a satisfactory Eckart embedding be defined even when $\mathbf{A}_c$ is considered.

Of the six permutations involving all four protons two, (1324) and (1423), yield singular $\mathbf{A}_p^P$ of the form

$$\begin{pmatrix} -4m_p a^2 & 0 \\ 0 & 0 \end{pmatrix}$$

and two, (1243) and (1342), yield singular $\mathbf{A}_p^P$ like

$$\begin{pmatrix} 0 & 0 \\ 0 & -4m_p b^2 \end{pmatrix}$$

This matrix can be rendered non-singular with the aid of $\mathbf{A}_c$. The permutation (1234) yields the non-singular but unsymmetric $\mathbf{A}_p^P$

$$\begin{pmatrix} 0 & 4m_p ab \\ -4m_p ab & 0 \end{pmatrix}$$

and (4321) yields the negative of this matrix.

Thus in the region of the reference geometry, of the 24 possible proton permutations, 16 lead to singular $\mathbf{A}_p^P$ and of these only 4 can lead to non-singular $\mathbf{A}^P$. The remaining 8 proton permutations form a group and all lead to non-singular $\mathbf{A}^P$ combined with either of the possible carbon permutations. This group has itself a sub-group of order 4 which gives rise to the perrotations discussed above.

3.2   Discussion and conclusions

What emerges from what has been said above is precisely how the imposition of a classical molecular geometry as a reference structure on a system in order to separate its internal motions from its rotations, breaks the permutational symmetry of the problem. It does so in such a way as to preclude a proper separation of rotations and internal motions if full permutational symmetrisation is attempted. That a molecular structure seems plausible within a quantum mechanical context, is undoubtedly due to the role played by the solutions of the clamped nuclei Hamiltonian in considering the separation of electronic and nuclear motion. Indeed just this point was made many years ago in the first section of a very interesting paper by Berry [7]. But it is perhaps worthwhile pointing out that, providing that the electronic coordinates are chosen relative to the centre of nuclear mass, there is absolutely no reason why the Eckart conditions should not be imposed on the nuclear part of the full problem, without any explicit separation of electronic motion. It is our classical preconceptions about a molecule that cause the trouble, rather than anything to do with the perceived virtues or limitations of that separation usually attributed to Born and Oppenheimer. This observation, if correct, must limit too the validity (though not, of course the utility) of the notion of *feasible* permutations, introduced by Longuet-Higgins [8] to deal with this problem and so much developed by later workers (see, for example [9]).

Since the problem is entirely of our own making it would seem foolish to persist in our error, particularly when there is an obvious way in which it can be avoided. If we choose the matrix **C** defining the rotational variables to be such as to diagonalise the instantaneous inertia tensor then all particles with the same mass will enter into the definition equivalently and permutational equivalence of the rotational variables will follow automatically. And indeed two of the first attempts to separate rotational from vibrational motion two, [10] and [11], adopted precisely this approach. It turns out that in this approach the definition of the frame embedded in the system depends upon the reciprocals of differences between instantaneous moments of inertia and therefore fails whenever two moments are equal. Thus the Hamiltonian could not be used to describe any symmetrical top molecule, for example ammonia. Indeed it turns out more generally, inspite of some heroic efforts by van Vleck [12], that the Hamiltonian so derived is largely ineffective in describing molecules in terms of their traditional geometrical structures and so it has found no use in the elucidation of molecular spectra.

It seems therefore that at present molecular structure has to be incorporated into the Hamiltonian in order to describe spectra effectively but that the price for this incorporation is that of a fundamental symmetry breaking.

Bibliography

[1] M. Born and J.R. Oppenheimer, *Ann.der Phys.*, 1927 **84**, 457.

[2] C. Eckart, *Phys. Rev.*, 1935, **47**, 552.

[3] J.K.G. Watson, *Mol. Phys.*, 1968, **15**, 479.

[4] R. J. Malhiot and S. M. Ferigle, *J.Chem. Phys*, **22**, 717-719, (1954).

[5] G. Ezra, *Symmetry properties of molecules* , Lecture Notes in Chemistry **28**, Springer-Verlag, Berlin, 1982.

[6] J. M. F. Gilles and J. Philippot, *Int. journ. quant. chem.*, 1972, **6**, 225.

[7] R. S. Berry, *Rev. Mod. Phys.*, 1960, **32**, 447.

[8] H. C. Longuet-Higgins, *Molec. Phys.*, 1963, **6**, 445.

[9] P. R. Bunker, *Molecular Symmetry and Spectroscopy*, Academic Press, London, 1979.

[10] C. Eckart, *Phys. Rev.*, 1934., **46**, 384.

[11] J. O. Hirschfelder and E. Wigner, *Proc. Nat. Acad. Sci.*, 1935, **21**, 113.

[12] J. H. van Vleck, *Phys. Rev.*, 1935, **47**, 487.

Chapter 2

SCF algorithms for HF electronic calculations

Eric Cancès
CERMICS, Ecole Nationale des Ponts et Chaussées,
6 & 8, avenue Blaise Pascal, Cité Descartes,
F-77455 Marne-La-Vallée Cedex, France
cances@cermics.enpc.fr

Abstract : This paper presents some mathematical results on SCF algorithms for solving the Hartree-Fock problem. In the first part of the article the focus is on two classical SCF procedures, namely the Roothaan algorithm and the level-shifting algorithm. It is demonstrated that the Roothaan algorithm either converges towards a solution to the Hartree-Fock equations or oscillates between two states which are not solution to the Hartree-Fock equations, any other behavior (oscillations between more than two states, "chaotic" behavior, ...) being excluded. The level-shifting algorithm is then proved to converge for large enough shift parameter, whatever the initial guess. The second part of the article details the convergence properties of a new algorithm recently introduced by Le Bris and the author, the so-called Optimal Damping Algorithm (ODA). Basic numerical simulations pointing out the principal features of the various algorithms under study are also provided.

1   Introduction

The Hartree-Fock (HF) model is a standard tool for computing an approximation of the ground state of a molecular system within the Born-Oppenheimer setting. From a mathematical viewpoint, the HF model gives rise to a nonquadratic constrained minimization problem for the numerical solution of which iterative procedures are needed; such procedures are referred to as Self-Consistent Field (SCF) algorithms. The solution to the HF problem can be obtained either by directly minimizing the HF energy functional [6, 11, 17, 25] or by solving the associated Euler-Lagrange equations, the so-called Hartree-Fock equations [20, 21, 22].

SCF algorithms for solving the HF equations are in general much more efficient than direct energy minimization techniques. However, these algorithms do not *a priori* ensure the decrease of the energy and they may lead to convergence problems [23]. For instance, the famous Roothaan algorithm (*see* [21] and section 4) is known to sometimes lead to stable oscillations between two states, none of them being a solution to the HF problem. This situation may occur even for simple chemical systems (see section 4).

Many articles have been devoted to the important issue of the SCF convergence. The behavior of the Roothaan algorithm is notably investigated in [2, 12] and in [26, 27]. In [2, 12] convergence difficulties are demonstrated for elementary two-dimensional models; in [26, 27], a stability condition of the Roothaan algorithm in the neighbourhood of a minimum of the HF energy is given for closed-shell systems. More sophisticated SCF algorithms for solving the HF equations have also been proposed to improve the convergence using various techniques like for instance damping [10, 28] or level-shifting [22]. Damping (as implemented in [28]) cures some convergence problems but many other remain. Numerical tests confirm that the level-shifting algorithm converges towards a solution to the HF equations for large enough shift parameters; a perturbation argument is provided in [22] to prove this convergence in the neighborhood of a stationary point. Unfortunately there is no guarantee that the so-obtained critical point of the HF energy functional is actually a minimum (even local); in addition, the level-shifting algorithm is known to only offer a slow speed of convergence. In practice, the most commonly used SCF algorithm is at the present time the Direct Inversion in the Iteration Space (DIIS) algorithm [20]. Numerical tests show that this algorithm is very efficient in most cases, but that it sometimes fails.

The present article belongs to a series of articles [3, 4] devoted to the SCF algorithms.

Our first purpose here is to report on recent mathematical results on the convergence properties of the Roothaan and of the level-shifting algorithms. Section 4 concerns the Roothaan algorithm, which is the most "natural" algorithm for solving the HF equations. Its is demonstrated that the Roothaan algorithm either converges towards a solution to the HF equations or oscillates between two states which are not solution to the HF equations, any other behavior being excluded. This theoretical result is in accordance with the numerical experiments. It is then explained in Section 5 why the introduction of a "level-shift" makes the algorithm converge.

The mathematical proofs are presented in the context of the *finite dimension* approximations of the HF problem obtained by a Galerkin method with a finite basis of atomic orbitals or plane waves, typically. They are consequently much simpler from a technical viewpoint than the proofs detailed in [3] which concern the original *infinite dimension* HF problem.

Recently, new SCF algorithms have been introduced in [4] by Le Bris and the author. They seem to exhibit good convergence properties at least for the chemical systems computed so far. These algorithms have been called Relaxed Constraints Algorithms (RCA) for they can be interpreted as direct minimization procedure of the HF energy which do not care about satisfying at each iteration the nonlinear constraints $D^2 = D$ that characterize admissible density matrices. The second purpose of this article (Section 6) is to detail the mathematical proof of the convergence of the basic RCA, namely the Optimal Damping Algorithm (ODA). Section 6 also contains some comments on the connections between RCA and other algorithms like the level-shifting and the DIIS algorithms.

Before coming up to our main topic, we devote Section 2 to a brief presentation of the HF model for readers (especially mathematicians) who are not familiar with Quantum Chemistry. Section 3 collects various general comments that apply to all the SCF algorithms considered in the sequel.

2   A brief presentation of the Hartree-Fock model

The problem under consideration consists in computing *ab initio*, that is to say without using any empirical parameter, the ground state energy of a molecular system made of M nuclei and N electrons. Tackling directly the $M + N$-body Schrödinger equation is today, and will probably remain, out of the scope of brute force numerical methods. Various approximations are therefore to be resorted to.

The first approximation that is common to most models of Quantum Chemistry is the so-called *Born-Oppenheimer approximation*. To make short, it consists in considering the nuclei as classical point particles. The Born-Oppenheimer approximation, which has been mathematically founded by Combes and al. [5], lays on the fact that nuclei are much heavier than electrons. The Born-Oppenheimer approximation is almost always valid in Chemistry (except for instance for studying specifically quantum phenomena involving nuclei as proton transfer by tunnel effect) and is therefore almost always used.

Within the Born-Oppenheimer approximation, the searching for the ground state takes the form of two nested minimization problems:

$$\inf \left\{ W(\bar{x}_1, \cdots, \bar{x}_M), \qquad (\bar{x}_1, \cdots, \bar{x}_M) \in \mathbb{R}^{3M} \right\} \qquad (1)$$

with

$$W(\bar{x}_1, \cdots, \bar{x}_M) = E_{el}(\bar{x}_1, \cdots, \bar{x}_M) + \sum_{1 \le k < l \le M} \frac{z_k \, z_l}{|\bar{x}_k - \bar{x}_l|}$$

$$E_{el}(\bar{x}_1, \cdots, \bar{x}_M) = \inf \left\{ \langle \psi, H_{\{\bar{x}_k\}} \psi \rangle, \quad \psi \in \mathcal{H}, \quad \|\psi\| = 1 \right\} \tag{2}$$

$$H_{\{\bar{x}_k\}} = -\sum_{i=1}^{N} \frac{1}{2} \Delta_{x_i} - \sum_{i=1}^{N} \sum_{k=1}^{M} \frac{z_k}{|x_i - \bar{x}_k|} + \sum_{1 \leq i < j \leq N} \frac{1}{|x_i - x_j|}$$

$$\mathcal{H} = \bigwedge_{i=1}^{N} L^2(\mathbb{R}^3 \times \{|+\rangle, |-\rangle\}, \mathbb{C})$$

In the above expressions, $\bar{x}_k$ denotes the current position in $\mathbb{R}^3$ of the k-th nucleus and z_k its charge.

The N electrons are described by a wave function $\psi(x_1, \sigma_1; \cdots; x_N, \sigma_N)$, where x_i and σ_i are respectively the position in $\mathbb{R}^3$ and the spin coordinate of the i-th electron. Each spin coordinate σ_i can take two values here denoted by $|+\rangle$ (spin up) and $|-\rangle$ (spin down). The wave function ψ is a normalized vector of the fermionic Hilbert space $\mathcal{H}$, and therefore satisfies on the one hand the antisymmetry condition

$$\psi(x_{p(1)}, \sigma_{p(1)}; \cdots; x_{p(N)}, \sigma_{p(N)}) = (-1)^{\epsilon(p)} \psi(x_1, \sigma_1; \cdots; x_N, \sigma_N)$$

for any permutation p of $\|[1, N]\|$ ($\epsilon(p)$ denoting the signature of p), and on the other hand the normalization condition

$$\sum_{\sigma_1, \cdots, \sigma_N} \int_{\mathbb{R}^{3N}} |\psi(x_1, \sigma_1; \cdots; x_N, \sigma_N)|^2 \, dx_1 \cdots dx_N = 1.$$

The operator $H_{\{\bar{x}_k\}}$ is the so-called electronic hamiltonian. It acts on $\mathcal{H}$; the $\bar{x}_k$ play the role of parameters. It is made of three terms, the first term accounting for the kinetic energy of the electrons, the second and the third terms accounting for nuclei-electrons and electrons-electrons interactions respectively. All physical quantities are expressed in atomic units [18].

Searching for the ground-state of the molecular system thus consists in minimizing the potential energy $W(\bar{x}_1, \cdots, \bar{x}_M)$ by solving the so-called *geometry optimization problem* (1). From the mathematical point of view, problem (1) is an unconstrained minimization problem of finite dimension. We refer the reader to [19, 24] for an overview of the various numerical methods dedicated to geometry optimization.

The specificity of problem (1) is that the function to be minimized, namely the potential energy W, is itself the result (up to the inter-nuclear repulsion term $\sum z_k z_l / |\bar{x}_k - \bar{x}_l|$) of the minimization problem (2) which is usually referred to as the *electronic problem*. We face this time a constrained minimization problem on the infinite dimension space $\mathcal{H}$.

In the sequel, we focus on the electronic problem, which is rewritten (in order to simplify the notations)

$$\inf \left\{ \langle \psi, H\psi \rangle, \quad \psi \in \mathcal{H}, \quad \|\psi\| = 1 \right\} \tag{3}$$

with

$$\mathcal{H} = \bigwedge_{i=1}^{N} L^2(\mathbb{R}^3 \times \{|+\rangle, |-\rangle\}, \mathbb{C})$$

$$H = -\sum_{i=1}^{N} \frac{1}{2}\Delta_{x_i} + \sum_{i=1}^{N} V(x_i) + \sum_{1 \le i < j \le N} \frac{1}{|x_i - x_j|}$$

$$V(x) = -\sum_{k=1}^{M} \frac{z_k}{|x - \bar{x}_k|}$$

the $\bar{x}_k$ being now fixed parameters in $\mathbb{R}^3$.

The Hartree-Fock approximation is of variational nature. It consists in restricting the set $\{\psi \in \mathcal{H}, \ \|\psi\| = 1\}$ on which the energy functional $\langle \psi, H\psi \rangle$ is minimized to the set of the Slater determinants, i.e. to the set of the wave functions ψ of the form

$$\psi = \frac{1}{\sqrt{N!}} \det(\phi_i(x_j, \sigma_j)) \tag{4}$$

where the ϕ_i, which are called *molecular orbitals*, satisfy the orthonormality conditions

$$\sum_{\sigma} \int_{\mathbb{R}^3} \phi_i(x, \sigma)\phi_j(x, \sigma)^* \, dx = \delta_{ij}.$$

A classical calculation (see [18] for instance) gives for any ψ of the form (4)

$$\langle \psi, H\psi \rangle = \mathcal{E}^{HF}(\{\phi_i\})$$

with

$$\mathcal{E}^{HF}(\{\phi_i\}) = \sum_{i=1}^{N} \frac{1}{2} \int_{\mathbb{R}^3} \sum_{\sigma} |\nabla\phi_i|^2 + \int_{\mathbb{R}^3} \rho_\Phi V$$

$$+ \frac{1}{2} \int_{\mathbb{R}^3} \int_{\mathbb{R}^3} \frac{\rho_\Phi(x)\,\rho_\Phi(x')}{|x - x'|} \, dx\, dx'$$

$$- \frac{1}{2} \int_{\mathbb{R}^3} \int_{\mathbb{R}^3} \sum_{\sigma,\sigma'} \frac{|\tau_\Phi(x, \sigma; x', \sigma')|^2}{|x - x'|} \, dx\, dx',$$

$$\tau_\Phi(x, \sigma; x', \sigma') = \sum_{i=1}^{N} \phi_i(x, \sigma)\, \phi_i(x', \sigma')^*,$$

$$\rho_\Phi(x) = \sum_{i=1}^{N} \sum_{\sigma} |\phi_i(x, \sigma)|^2.$$

The HF problem thus reads

$$\inf \left\{ \mathcal{E}^{HF}(\{\phi_i\}), \quad \phi_i \in L^2(\mathbb{R}^3 \times \{|+\rangle, |-\rangle\}, \mathbb{C}), \quad \sum_{\sigma} \int_{\mathbb{R}^3} \phi_i(x, \sigma)\phi_j(x, \sigma)^* \, dx = \delta_{ij} \right\}.$$

The mathematical properties of the HF problem have been studied by Lieb and Simon [14] and by P.-L. Lions [15]. The existence of a HF electronic ground state is guaranteed for positive ions ($Z := \sum_{k=1}^{M} z_k > N$) and neutral systems ($Z = N$). We are not aware of any general existence result for negative ions (the available existence proofs only work for $N < Z+1$). On the other hand, there is a non-existence results for negative ions such that $N > 2Z + M$ [13] (this inequality holds for instance for the ion H^{2-}). As far as we know, uniqueness (of the density ρ at least) is an open problem, probably of outstanding difficulty.

The last step of the approximation procedure consists in approaching the *infinite dimensional* HF problem by a *finite dimensional* HF problem by means of a Galerkin approximation: the HF energy is minimized over the set of molecular orbitals that can be expanded on a given finite basis $\{\chi_p\}_{1 \le p \le n}$:

$$\phi_i = \sum_{k=1}^{n} C_{ki} \chi_k.$$

Denoting by $S = [S_{kl}]$ with

$$S_{kl} = \sum_{\sigma} \int_{\mathbf{R}_3} \chi_k^* \chi_l$$

the so-called *overlap* matrix, the constraints $\sum_{\sigma} \int_{\mathbf{R}^3} \phi_i \phi_j^* = \delta_{ij}$ read

$$\delta_{ij} = \sum_{\sigma} \int_{\mathbf{R}^3} \phi_i \phi_j^* = \sum_{\sigma} \int_{\mathbf{R}^3} \left(\sum_{l=1}^{n} C_{li} \chi_l \right) \left(\sum_{k=1}^{n} C_{kj} \chi_k \right)^* = \sum_{k=1}^{n} \sum_{l=1}^{n} C_{kj}^* S_{kl} C_{li},$$

or in matrix form

$$C^* S C = I_N,$$

where I_N denotes the identity matrix of rank N. In addition,

$$
\begin{aligned}
\sum_{i=1}^{N} \frac{1}{2} \int_{\mathbf{R}^3} |\nabla \phi_i|^2 + \int_{\mathbf{R}^3} \rho_{\Phi} V &= \sum_{i=1}^{N} \left(\frac{1}{2} \int_{\mathbf{R}^3} |\nabla \phi_i|^2 + \int_{\mathbf{R}^3} V |\phi_i|^2 \right) \\
&= \sum_{i=1}^{N} \left(\frac{1}{2} \int_{\mathbf{R}^3} \left| \nabla \sum_{k=1}^{n} C_{ik} \chi_k \right|^2 + \int_{\mathbf{R}^3} V \left| \sum_{k=1}^{n} C_{ik} \chi_k \right|^2 \right) \\
&= \sum_{i=1}^{N} \sum_{k=1}^{n} \sum_{l=1}^{n} h_{kl} C_{li} C_{ki}^* \\
&= \text{Tr} \left(h C C^* \right)
\end{aligned}
$$

where h denotes the matrix of the core hamiltonian $-\frac{1}{2}\Delta + V$ in the basis $\{\chi_k\}$:

$$h_{kl} = \frac{1}{2} \sum_{\sigma} \int_{\mathbf{R}_3} \nabla \chi_k^* \cdot \nabla \chi_l + \sum_{\sigma} \int_{\mathbf{R}^3} V \chi_k^* \chi_l.$$

Lastly, denoting by

$$(ij|kl) = \sum_{\sigma} \sum_{\sigma'} \int_{\mathbf{R}^3} \int_{\mathbf{R}^3} \frac{\chi_i(x)\chi_j(x)^* \chi_k(x')\chi_l(x')^*}{|x - x'|} \, dx \, dx' \quad \text{and} \quad A_{ijkl} = (ij|kl) - (il|kj),$$

the inter-electronic repulsion term reads

$$\int_{\mathbf{R}^3} \int_{\mathbf{R}^3} \frac{\rho_{\Phi}(x) \, \rho_{\Phi}(x')}{|x - x'|} \, dx \, dx' - \int_{\mathbf{R}^3} \int_{\mathbf{R}^3} \frac{|\tau_{\Phi}(x, x')|^2}{|x - x'|} \, dx \, dx' = \sum_{i,j,k,l=1}^{n} \sum_{\alpha,\beta=1}^{N} A_{ijkl} C_{i\alpha} C_{j\alpha}^* C_{k\beta} C_{l\beta}^*.$$

The above expressions incite one to introduce the so-called *density matrix*

$$D = CC^*,$$

which permits to write the HF energy under the compact form

$$E^{HF}(D) = \text{Tr}\,(hD) + \frac{1}{2}\text{Tr}\,(G(D)D),$$

where $G(D)$ denotes the contracted product of the 4-index tensor A by D:

$$G(D)_{ij} = (A : D)_{ij} = \sum_{kl} A_{ijkl}D_{kl}.$$

It is easy to see that the matrices D which read $D = CC^*$ with $C \in \mathcal{M}(n, N)$ and $C^*SC = I_N$ are those which satisfy $\text{Tr}\,(SD) = N$ and $DSD = D$. The so-obtained finite dimension HF problem then reads

$$\inf\left\{E^{HF}(D), \quad D \in \mathcal{M}(n, n), \quad D^* = D, \quad \text{Tr}\,(SD) = N, \quad DSD = D\right\}.$$

For the sake of simplicity, we assume in the sequel that the overlap matrix S equals identity, that is to say that the basis $\{\chi_p\}_{1 \le p \le n}$ is orthonormal. The general case is recovered by the transformation rules $D \to S^{1/2}DS^{1/2}$, $h \to S^{-1/2}hS^{-1/2}$, $G(D) \to S^{-1/2}G(S^{1/2}DS^{1/2})S^{-1/2}$. The HF problem then reads:

$$\inf\left\{E^{HF}(D), \quad D \in \mathcal{P}\right\}, \tag{5}$$

with

$$\mathcal{P} = \left\{D \in \mathcal{M}(n, n), \quad D^* = D, \quad \text{Tr}\,D = N, \quad D^2 = D\right\}.$$

The following lemma provides a characterization of the critical points of the HF minimization problem (5).

Lemma 1. *For any D, let us denote by $F(D) = h + G(D)$ the Fock matrix associated with D.*

1. *A density matrix $D \in \mathcal{P}$ is a critical point of the HF problem (5) if and only if*

$$\begin{cases} F(D)C = CE \\ C^*C = I_N \\ D = CC^* \end{cases} \tag{6}$$

where $E = Diag(\epsilon_1, \epsilon_2, \cdots, \epsilon_N)$ is a $N \times N$ diagonal matrix collecting N eigenvalues of the linear eigenvalue problem

$$F(D) \cdot \phi = \epsilon\,\phi.$$

and where C is a $n \times N$ matrix containing N orthonormal eigenvectors associated with ϵ_1, ϵ_2, ..., ϵ_N. The condition (6) is equivalent to the condition

$$[F(D), D] = 0,$$

where $[\cdot, \cdot]$ denotes the matrix commutator defined for any A and B in $\mathcal{M}(n, n)$ by $[A, B] = AB - BA$.

2. *For $D \in \mathcal{P}$ being a local minimum of the HF problem, it is necessary that ϵ_1, ϵ_2, ..., ϵ_N are the smallest N eigenvalues of $F(D)$ including multiplicity.*

This result is classical; its proof can be read in any textbook of Quantum Chemistry (see [18] for instance).

Remark. The model described above is the so-called General Hartree-Fock (GHF) model. Most often in practice, Quantum Chemistry calculations are performed with spin constraints models like the Restricted Hartree-Fock (RHF), the Unrestricted Hartree-Fock (UHF) or the Restricted Open-shell Hartree-Fock (ROHF) models. The convergence results stated below can be adapted without difficulties to these models. ◊

3   General remarks on SCF algorithms

Various SCF algorithms are studied in the following three sections. All of them consist in generating a sequence (D_k) defined by

$$\begin{cases} \tilde{F}_k C_{k+1} = C_{k+1} E_{k+1} \\ C_{k+1}^* C_{k+1} = I_N \\ D_{k+1} = C_{k+1} C_{k+1}^* \end{cases} \tag{7}$$

where $E_{k+1} = \mathrm{Diag}(\epsilon_1^{k+1}, \cdots, \epsilon_N^{k+1})$, $\epsilon_1^{k+1} \leq \epsilon_2^{k+1} \leq \cdots \leq \epsilon_n^{k+1}$ being the eigenvalues of the linear eigenvalue problem

$$\tilde{F}_k \cdot \phi = \epsilon \, \phi,$$

and where C_{k+1} collects N orthonormal eigenvectors associated with ϵ_1^{k+1}, ϵ_2^{k+1}, ..., ϵ_N^{k+1}. The expression of the current Fock matrix $\tilde{F}_k$ characterizes the algorithm. We have for instance

- $\tilde{F}_k = F(D_k)$ for the Roothaan algorithm;

- $\tilde{F}_k = F(D_k) - bD_k$ where b is a positive constant for the level-shifting algorithm;

- $\tilde{F}_k = F(\widetilde{D}_k)$ for the ODA, where $\widetilde{D}_k$ is a pseudo-density matrix which satisfies the *relaxed* constraints $\widetilde{D}_k^2 \leq \widetilde{D}_k$ and is defined so that the HF energy $E^{HF}(\widetilde{D}_k)$ decreases at each iteration (see section 6).

The procedure consisting in assembling the matrix $D_{k+1} \in \mathcal{P}$ by populating the N molecular orbitals of lowest energies of the current Fock matrix $\tilde{F}_k$ is referred to as the *aufbau principle*. It is justified by the results stated in Lemma 1. For the matrix D_{k+1} being defined in a unique way, it suffices that $\epsilon_N^{k+1} < \epsilon_{N+1}^{k+1}$. Degeneracies in the spectrum are in general related to the symmetries of the system: in the cases when the system does not exhibit any symmetry, numerical experiments show that the eigenvalues of $\tilde{F}_k$ are generically non-degenerate for any k, whereas it may not be the case when the system does exhibit symmetries (consider for instance the spherical symmetry of the hamiltonian in the atomic case). Degeneracies create technical difficulties which complicate the theoretical studies on SCF convergence. For the sake of simplicity, we therefore assume from now on that the *uniform well-posedness* (UWP) property introduced in [3] is satisfied:

UWP property: a SCF algorithm of the form (7) with initial guess D_0 will be said to be uniformly well-posed if there exists some positive constant γ such that

$$\epsilon_{N+1}^{k+1} \geq \epsilon_N^{k+1} + \gamma.$$

The consequences of the UWP assumption which will be useful below have been collected in the following lemma, whose proof is postponed until the end of the present section.

Lemma 2. *Let us consider a SCF algorithm of the form (7) with initial guess D_0 which satisfies the UWP property. Then*

1. *The updated density matrix D_{k+1} is defined in a unique way at each iteration; this matrix can be characterized as the minimizer of the variational problem*

$$\inf \left\{ \mathrm{Tr}\,(\tilde{F}_k D), \quad D \in \mathcal{P} \right\}.$$

2. *For any $D \in \mathcal{M}(n,n)$ such that $D = D^*$, $\mathrm{Tr}\,(D) = N$ and $D^2 \leq D$,*

$$\mathrm{Tr}\,(\tilde{F}_k D) \geq \mathrm{Tr}\,(\tilde{F}_k D_{k+1}) + \frac{\gamma}{2}\|D - D_{k+1}\|^2,$$

$\|\cdot\|$ *denoting the Hilbert-Schmidt norm defined for any $A \in \mathcal{M}(n,n)$ by $\|A\| = \mathrm{Tr}\,(AA^*)^{1/2}$.*

In the sequel, we denote by arg inf $\mathcal{MP}$ the minimizer of the minimization problem $\mathcal{MP}$. We can therefore write

$$D_{k+1} = \arg\inf \left\{ \mathrm{Tr}\,(\tilde{F}_k D), \quad D \in \mathcal{P} \right\}.$$

Remark. Let us point out that some convergence results can be obtained without resorting to the UWP assumption. In particular, it turns out that the level-shifting algorithm is automatically UWP as soon as the shift parameter is large enough (see [3] for details). It can also be proved that the ODA numerically converges towards an *aufbau* solution to the HF equations *within the GHF setting* and provided the basis is "large enough". We do not detail here the rather technical proof of this assertion. Let us just mention that it is based on a mathematical result by P.-L. Lions [16] related to finite-temperature HF models. Unfortunately, so far as we know, the arguments used in [16] cannot be extended to the RHF, UHF or ROHF models. ◇

Before turning to the study of SCF algorithms, the notion of convergence has to be made precise. We are in fact not able to prove mathematical convergence results of the form "the sequence (D_k) converges towards a minimizer D of the HF problem (5)" for at least two reasons. First, we are solving the Euler-Lagrange equations associated with the HF minimization problem (5), namely the HF equations (6); even in case of convergence we have no argument to conclude that the so-obtained critical point is actually a minimum (even local) of the HF energy. Second, we have no precise description of the topology of the set of the critical points of (5); this lack of information prevents us from proving the convergence of the whole sequence (D_k) towards a solution D to the HF equations. We can at best obtain that $D_{k+1} - D_k$ goes to zero, and that for "large" k, D_k is "close to" a solution to the HF equations (6) satisfying the *aufbau* principle. For instance, it may happen that the HF problem admits a connected manifold of minima; this phenomenon is observed in particular for open-shell atoms because the spherical symmetry of the problem is broken by the HF approximation (this can be related to a mathematical result by Bach, Lieb, Loss and Solovej [1] stating that "there are no unfilled shell" in the HF ground states). We cannot then discriminate between the case when the sequence (D_k) converges towards a point of the manifold and the case when the sequence (D_k) is attracted by the manifold together with a slow drift parallel to the manifold.

We shall consequently adopt here the following two convergence criteria, which are sufficient in practice. We shall say that a SCF algorithm of the form (7) *numerically converges* towards a solution to the HF equations if the sequence (D_k) satisfies

1. $D_{k+1} - D_k \longrightarrow 0$;

2. $[F(D_k), D_k] \longrightarrow 0$;

and that it *numerically converges* towards an *aufbau* solution to the HF equations if the sequence (D_k) satisfies

1. $D_{k+1} - D_k \longrightarrow 0$;

2. $\mathrm{Tr}\,(F(D_k)D_k) - \inf\{\mathrm{Tr}\,(F(D_k)D),\quad D \in \mathcal{P}\} \longrightarrow 0$.

As all norms are equivalent in finite dimension, we do not need to specify the matrix norm in which the variations are evaluated. Let us remark that the latter convergence criterion is stronger than the former one for

$$(\mathrm{Tr}\,(F(D_k)D_k) - \inf\{\mathrm{Tr}\,(F(D_k)D),\quad D \in \mathcal{P}\} \to 0) \quad \Rightarrow ([F(D_k), D_k] \to 0).$$

Let us conclude this section with the

Proof of Lemma 2. Let us denote by D a current matrix such that $D = D^*$, $\mathrm{Tr}\,(D) = N$, $D^2 \leq D$ and by D_{ij} its coefficients in an orthonormal basis in which $\tilde{F}_k = \mathrm{Diag}(\epsilon_1^{k+1}, \epsilon_2^{k+1}, \cdots, \epsilon_n^{k+1})$ with $\epsilon_1^{k+1} \leq \epsilon_2^{k+1} \leq \cdots \leq \epsilon_n^{k+1}$. In such a basis $D_{k+1} = \mathrm{Diag}(1, \cdots, 1, 0, \cdots, 0)$. As in addition $\mathrm{Tr}\,(D) = \sum_{i=1}^{n} D_{ii} = N$, we get first

$$\begin{aligned}
\|D_{k+1} - D\|^2 &= \mathrm{Tr}\,((D_{k+1} - D) \cdot (D_{k+1} - D)) \\
&= \mathrm{Tr}\,(D_{k+1}^2) + \mathrm{Tr}\,(D^2) - 2\,\mathrm{Tr}\,(DD_{k+1})
\end{aligned}$$

$$\leq \text{ Tr } (D_{k+1}) + \text{Tr } (D) - 2\,\text{Tr } (DD_{k+1})$$

$$= 2N - 2\sum_{i=1}^{N} D_{ii}$$

$$= 2 \sum_{i=N+1}^{n} D_{ii}.$$

Besides $\text{Tr } (\tilde{F}_k D_{k+1}) = \sum_{i=1}^{N} \epsilon_i^{k+1}$, $\text{Tr } (\tilde{F}_k D) = \sum_{i=1}^{n} \epsilon_i^{k+1} D_{ii}$ and

$$0 \leq D_{ii} \leq 1, \qquad \text{for any } 1 \leq i \leq n$$

for $D^2 \leq D = D^*$ implies $|D_{ii}|^2 + \sum_{j \neq i} |D_{ij}|^2 \leq D_{ii}$. Putting together the above results, we obtain

$$\begin{aligned}
\text{Tr } (\tilde{F}_k D) &= \sum_{i=1}^{n} \epsilon_i^{k+1} D_{ii} \\
&\geq \sum_{i=1}^{N} \epsilon_i^{k+1} D_{ii} + \sum_{i=N+1}^{n} (\epsilon_N^{k+1} + \gamma) D_{ii} \\
&= \sum_{i=1}^{N} \epsilon_i^{k+1} D_{ii} + \epsilon_N^{k+1} \sum_{i=N+1}^{n} D_{ii} + \gamma \sum_{i=N+1}^{n} D_{ii} \\
&= \sum_{i=1}^{N} \epsilon_i^{k+1} D_{ii} + \epsilon_N^{k+1} (N - \sum_{i=1}^{N} D_{ii}) + \gamma \sum_{i=N+1}^{n} D_{ii} \\
&= \sum_{i=1}^{N} \epsilon_i^{k+1} + \sum_{i=1}^{N} (\epsilon_N^{k+1} - \epsilon_i^{k+1})(1 - D_{ii}) + \gamma \sum_{i=N+1}^{n} D_{ii}.
\end{aligned}$$

As for any $1 \leq i \leq N$, $0 \leq D_{ii} \leq 1$ and $\epsilon_N^{k+1} \geq \epsilon_i^{k+1}$, we finally obtain

$$\text{Tr } (\tilde{F}_k D) \geq \text{Tr } (\tilde{F}_k D_{k+1}) + \frac{\gamma}{2} \|D - D_{k+1}\|^2.$$

The two statements of Lemma 2 follow. $\diamond$

4 The Roothaan algorithm: why and how it fails

The Roothaan algorithm (also called *simple SCF* or *pure SCF* or *conventional SCF* in the literature) is the simplest fixed point procedure associated with the nonlinear eigenvalue problem (6). It consists in generating a sequence (D_k^{Rth}) in $\mathcal{P}$ satisfying

$$\begin{cases} F(D_k^{Rth}) C_{k+1} = C_{k+1} E_{k+1} \\ C_{k+1}^* C_{k+1} = I_N \\ D_{k+1}^{Rth} = C_{k+1} C_{k+1}^* \end{cases}$$

where $E_{k+1} = \text{Diag}(\epsilon_1^{k+1}, \cdots, \epsilon_N^{k+1})$, $\epsilon_1^{k+1} \leq \epsilon_2^{k+1} \leq \cdots \leq \epsilon_N^{k+1}$ being the N smallest eigenvalues of the linear eigenvalue problem

$$F(D_k^{Rth}) \cdot \phi = \epsilon\, \phi$$

and where the $n \times N$ matrix C_{k+1} collects N orthonormal eigenvectors of $F(D_k^{Rth})$ associated with ϵ_1^{k+1}, ϵ_2^{k+1}, ..., ϵ_N^{k+1}. The iteration procedure of the Roothaan algorithm can therefore be summarized by the diagram

$$D_k^{Rth} \longrightarrow \tilde{F}_k = F(D_k^{Rth}) \overset{\text{aufbau}}{\longrightarrow} D_{k+1}^{Rth}.$$

The convergence properties of the Roothaan algorithm are not satisfactory: although the Roothaan algorithm sometimes numerically converges towards a solution to the HF equations, it frequently numerically oscillates between two states, none of them being solution to the HF equations. Numerical oscillation between two states means here that

$$D_{k+2}^{Rth} - D_k^{Rth} \longrightarrow 0, \quad \text{but} \quad D_{k+1}^{Rth} - D_k^{Rth} \not\longrightarrow 0.$$

The behavior of the Roothaan algorithm can be explained by introducing the auxiliary function

$$E(D, D') = \text{Tr}\,(hD) + \text{Tr}\,(hD') + \text{Tr}\,(G(D)\,D'),$$

which is symmetric since $\text{Tr}\,(G(D)\,D') = \text{Tr}\,(G(D')\,D)$, and which satisfies $E(D, D) = 2\,E^{HF}(D)$. Let us indeed minimize E alternatively with respect to each of the two arguments D and D':

$$D_1 = \arg\inf\{E(D_0, D), \quad D \in \mathcal{P}\},$$

$$D_2 = \arg\inf\{E(D, D_1), \quad D \in \mathcal{P}\},$$

$$D_3 = \arg\inf\{E(D_2, D), \quad D \in \mathcal{P}\},$$

$$\ldots$$

This minimization procedure is usually called *relaxation* in the mathematical literature. For the first two steps, we obtain

$$\begin{aligned}
D_1 &= \arg\inf\{E(D_0, D), \quad D \in \mathcal{P}\} \\
&= \arg\inf\{\text{Tr}\,(hD_0) + \text{Tr}\,(hD) + \text{Tr}\,(G(D_0)D), \quad D \in \mathcal{P}\} \\
&= \arg\inf\{\text{Tr}\,(F(D_0)D), \quad D \in \mathcal{P}\} \\
&= D_1^{Rth},
\end{aligned}$$

and, since E is symmetric on $\mathcal{P} \times \mathcal{P}$,

$$\begin{aligned}
D_2 &= \arg\inf\{E(D, D_1^{Rth}), \quad D \in \mathcal{P}\} \\
&= \arg\inf\{E(D_1^{Rth}, D), \quad D \in \mathcal{P}\} \\
&= \arg\inf\{\text{Tr}\,(hD) + \text{Tr}\,(hD_1^{Rth}) + \text{Tr}\,(G(D_1^{Rth})D), \quad D \in \mathcal{P}\} \\
&= \arg\inf\{\text{Tr}\,(F(D_1^{Rth})D), \quad D \in \mathcal{P}\} \\
&= D_2^{Rth}.
\end{aligned}$$

It follows by induction that the sequences generated by the relaxation algorithm on the one hand, and by the Roothaan algorithm on the other hand, are the same. The functional E, which decreases at each iteration of the relaxation procedure can therefore be interpreted as a Lyapunov functional of the Roothaan algorithm. This basic remark is the foundation of the proof of the following result.

Theorem 1. *Let $D_0 \in \mathcal{P}$ such that the Roothaan algorithm with initial guess D_0 is UWP. Then the sequence (D_k^{Rth}) generated by the Roothaan algorithm satisfies one of the following two properties*

- *either (D_k^{Rth}) numerically converges towards an aufbau solution to the HF equations*

- *or (D_k^{Rth}) numerically oscillates between two states, none of them being an aufbau solution to the HF equations.*

Proof. For any $k \in \mathbb{N}$, we deduce from Lemma 2 that

$$\mathrm{Tr}\,(F(D_{k+1}^{Rth})D_{k+2}^{Rth}) + \frac{\gamma}{2}\|D_{k+2}^{Rth} - D_k^{Rth}\|^2 \leq \mathrm{Tr}\,(F(D_{k+1}^{Rth})D_k^{Rth}).$$

Adding $\mathrm{Tr}\,(hD_{k+1}^{Rth})$ to both terms of the above inequality, we obtain

$$E(D_{k+1}^{Rth}, D_{k+2}^{Rth}) + \frac{\gamma}{2}\|D_{k+2}^{Rth} - D_k^{Rth}\|^2 \leq E(D_k^{Rth}, D_{k+1}^{Rth}).$$

We then sum up the above inequalities for $k \in \mathbb{N}$ and we get $\sum_{k \in \mathbb{N}}\|D_{k+2}^{Rth} - D_k^{Rth}\|^2 < +\infty$, which involves in particular that

$$D_{k+2}^{Rth} - D_k^{Rth} \longrightarrow 0.$$

Now, either $D_{k+1}^{Rth} - D_k^{Rth}$ converges to *zero* or it does not. In the former case, we deduce from the characterization of D_{k+1}^{Rth} by

$$\mathrm{Tr}\,(F(D_k^{Rth})D_{k+1}^{Rth}) = \inf\left\{\mathrm{Tr}\,(F(D_k^{Rth})D), \quad D \in \mathcal{P}\right\}$$

that

$$\mathrm{Tr}\,(F(D_k^{Rth})D_k^{Rth}) - \inf\left\{\mathrm{Tr}\,(F(D_k^{Rth})D), \quad D \in \mathcal{P}\right\} \longrightarrow 0.$$

Convergence towards an *aufbau* solution to the HF equations is thus established. In the latter case

$$\mathrm{Tr}\,(F(D_{2k}^{Rth})D_{2k}^{Rth}) - \inf\left\{\mathrm{Tr}\,(F(D_{2k}^{Rth})D), \quad D \in \mathcal{P}\right\}$$
$$= \mathrm{Tr}\,(F(D_{2k}^{Rth})D_{2k}^{Rth}) - \mathrm{Tr}\,(F(D_{2k}^{Rth})D_{2k+1}^{Rth})$$
$$\geq \frac{\gamma}{2}\|D_{2k}^{Rth} - D_{2k+1}^{Rth}\|^2 \not\longrightarrow 0.$$

Convergence of (D_{2k}) towards an *aufbau* solution to the HF equations is therefore excluded; the same argument holds for (D_{2k+1}). ◇

Mimicking the proof of Theorem 2 (see section 5), it is easy to establish in addition that (D_{2k}, D_{2k+1}) converges up to an extraction to a critical point $(D, D') \in \mathcal{P} \times \mathcal{P}$ of the functional E which satisfies

$$
\begin{cases}
F(D')C = CE \\
C^*C = I_N \\
D = CC^* \\
F(D)C' = C'E' \\
C'^*C' = I_N \\
D' = C'C'^*
\end{cases}
$$

where E and E' are diagonal matrices collecting the smallest N eigenvalues of $F(D')$ and $F(D)$ respectively. Besides, as $E(D_{2k}, D_{2k+1})$ is decreasing, the whole sequence (D_{2k}, D_{2k+1}) converges to (D, D') if this critical point is a strict (local) minimum. In this case, the alternatives are

- either (D, D') is on the diagonal of $\mathcal{P} \times \mathcal{P}$ (i.e. $D = D'$) and (D_k^{Rth}) converges towards an *aufbau* solution to the HF equations;

- or (D, D') is not on the diagonal of $\mathcal{P} \times \mathcal{P}$ (i.e. $D \neq D'$) and (D_k^{Rth}) oscillates between two states which are not *aufbau* solutions to the HF equations.

Both situations are represented on Figure 2.1.

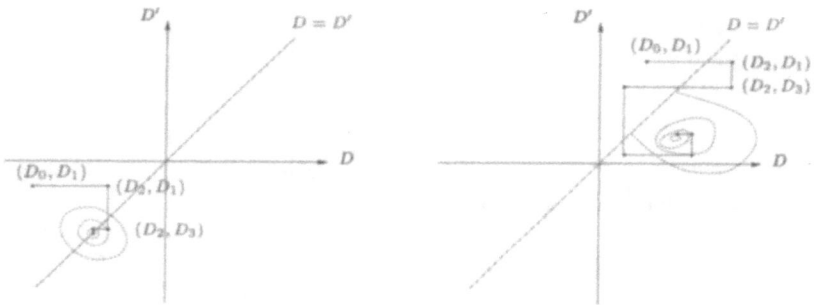

Figure 2.1: Minimization of E by relaxation: convergence towards a strict local minimum located on (resp. off) the "diagonal" leads to the convergence (resp. oscillations) of the Roothaan algorithm.

Oscillations can be observed even for simple chemical systems. As a matter of example, we have tested the Roothaan algorithm in the UHF setting for the atoms of the periodic table and for two sets of atomic orbitals, namely the Gaussian basis sets 3-21G and 6-311++G(3df,3pd) (see [9]). The initial guess is obtained by diagonalization of the core hamiltonian. Calculations have been performed with Gaussian 98 [8]. The results are reported in Figure 2.2; they indicate that

1. Both alternatives (convergence *vs* oscillation) are met in practice.

2. Convergence towards a critical point of the HF problem which is not a global minimum can sometimes be observed.

3. For the same system, we can get convergence for one basis set and oscillation for another basis set.

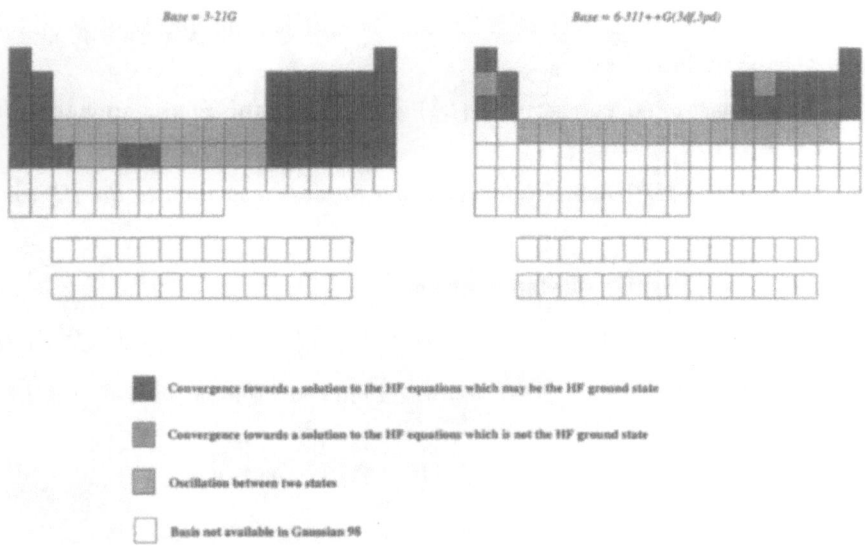

Figure 2.2: Searching the ground state of atoms with the Roothaan algorithm. Results are shown on the periodic table of the elements.

5   Level-shifting

The analysis developed in the previous section suggests to add to the functional E a penalty term E_p of the off-diagonal pairs (D, D') with $D \neq D'$ in order to enforce the critical points of the functional $E + E_p$ to lie on the diagonal of $\mathcal{P} \times \mathcal{P}$, which should ensure convergence towards a critical point of (5).

A simple penalty fonctional reads $E_p = b\|D - D'\|^2$, where b is a positive constant and where $\|\cdot\|$ denotes as above the Hilbert-Schmidt norm. Let us therefore set

$$E^b(D, D') = \mathrm{Tr}\,(hD) + \mathrm{Tr}\,(hD') + \mathrm{Tr}\,(G(D)\,D') + b\,\|D - D'\|^2.$$

The relaxation algorithm associated with the minimization problem

$$\inf\left\{E^b(D, D'), \quad (D, D') \in \mathcal{P} \times \mathcal{P}\right\}$$

generates the sequence (D_k^b) defined by

$$D_k^b \longrightarrow \tilde{F}_k = F(D_k^b) - b\,D_k^b \xrightarrow{\text{aufbau}} D_{k+1}^b.$$

The sequence (D_k^b) can be identified with the sequence generated by the so-called level-shifting algorithm [22] with level-shift parameter b. The convergence of the level-shifting algorithm towards a (non necessarily *aufbau*) solution to the HF equations is mathematically guaranteed:

Theorem 2. *There exists a positive constant b_0 such that for any $D_0 \in \mathcal{P}$ and for any level-shift parameter $b \geq b_0$,*

1. *The sequence of the energies $E^{HF}(D_n^b)$ decreases towards some stationary value $\mathcal{E}$ of E^{HF}.*

2. *The sequence (D_n^b) numerically converges towards a solution to the HF equations.*

Proof. Let b_0 be a positive constant such that

$$\forall(D, D') \in \mathcal{M}(n,n) \times \mathcal{M}(n,n), \quad \mathrm{Tr}\ (G(D - D') \cdot (D - D')) \leq b_0 \|D - D'\|^2. \quad (8)$$

Such a b_0 exists since $(d, d') \mapsto \mathrm{Tr}\ (G(d)d')$ is a bilinear form on $\mathcal{M}(n,n)$. As E^b is symmetric on $\mathcal{P} \times \mathcal{P}$, we have for any $k \in \mathbb{N}$,

$$E^b(D_k^b, D_{k+1}^b) = \inf \left\{ E^b(D_k^b, D), \quad D \in \mathcal{P} \right\}$$
$$\leq E^b(D_k^b, D_k^b).$$

A simple calculation shows that this inequality can be rewritten as

$$E^{HF}(D_{k+1}^b) - \frac{1}{2}\mathrm{Tr}\ (G(D_{k+1}^b - D_k^b) \cdot (D_{k+1}^b - D_k^b)) + b\|D_{k+1}^b - D_k^b\|^2 \leq E^{HF}(D_k^b).$$

Therefore, for any $b \geq b_0$,

$$E^{HF}(D_{k+1}^b) + \frac{b}{2}\|D_{k+1}^b - D_k^b\|^2 \leq E^{HF}(D_k^b). \quad (9)$$

It follows that $(E^{HF}(D_k^b))$ is a decreasing sequence and that

$$\sum_{k=0}^{+\infty} \|D_{k+1}^b - D_k^b\|^2 < +\infty.$$

The latter statement, which has been obtained by summing the inequalities (9) for $k \geq 0$, implies in particular that

$$D_{k+1}^b - D_k^b \longrightarrow 0.$$

As for any $k \in \mathbb{N}$,

$$[F(D_k^b) - bD_k^b, D_{k+1}^b] = 0,$$

it follows that

$$[F(D_k^b), D_k^b] = [F(D_k^b) - bD_k^b, D_{k+1}^b - D_k^b] \xrightarrow[k \to +\infty]{} 0.$$

This concludes the proof of statement 2. As $\mathcal{P}$ is compact, we can extract from $(D_k^b)_{k \in \mathbb{N}}$ a subsequence $(D_{k_l}^b)_{l \in \mathbb{N}}$ which converges towards some $D \in \mathcal{P}$, such that $E^{HF}(D_{k_l}^b) \downarrow E^{HF}(D)$ and $[F(D), D] = 0$; $\mathcal{E} = \lim E^{HF}(D_k) = E^{HF}(D)$ is therefore a stationary value of the HF energy. $\Diamond$

The level-shift parameter b_0 implicitly defined by (8) is far from being optimal. Explicit and more refined estimates of shift parameters that guarantee convergence are given in [3]. From a numerical viewpoint, it is important to choose not too large a shift parameter; otherwise the speed of convergence is very slow and the risk of converging towards a critical point whose energy is above that of the HF ground state is enhanced. This point, on which we will come back in the course of section 6, is illustrated by the numerical example reported on Figure 2.3 in which the level-shifting algorithm with various shift parameters has been used to compute the (doublet) UHF ground state of the Bromine atom in the Gaussian basis 6-311++G(3df,3pd) (see [9]). In each case, the initial guess is computed by diagonalizing the core hamiltonian. Calculations have been performed with Gaussian 98 [8]. The algorithm oscillates for small shift parameters. For larger shift parameters, damped oscillations leading to convergence are observed. For very large shift parameter, the energy decreases at each iteration. Too large shift parameters have however to be excluded because they slow down the convergence (for $b = 30.0$ Ha, convergence towards the ground state is obtained after more than 200 iterations).

6   The Optimal Damping Algorithm

The present section is devoted to the mathematical study of the Optimal Damping Algorithm (ODA) which is the simplest representative of the class of Relaxed Constraints Algorithms (RCA) introduced in [4].

The ODA is defined by the following two-step iteration procedure

1. Diagonalize the current Fock matrix $\widetilde{F}_k = F(\widetilde{D}_k)$ and assemble the matrix $D_{k+1} \in \mathcal{P}$ by the *aufbau* principle;

2. Set $\widetilde{D}_{k+1} = \arg\inf\left\{E(\widetilde{D}), \quad \widetilde{D} \in \mathrm{Seg}[\widetilde{D}_k, D_{k+1}]\right\}$ where

$$\mathrm{Seg}[\widetilde{D}_k, D_{k+1}] = \left\{(1 - \lambda)\widetilde{D}_k + \lambda D_{k+1}, \quad \lambda \in [0, 1]\right\}$$

denotes the line segment linking $\widetilde{D}_k$ and D_{k+1}.

The procedure is initialized with $\widetilde{D}_0 = D_0$, $D_0 \in \mathcal{P}$ being a given initial guess.

The ODA thus generates two sequences of matrices:

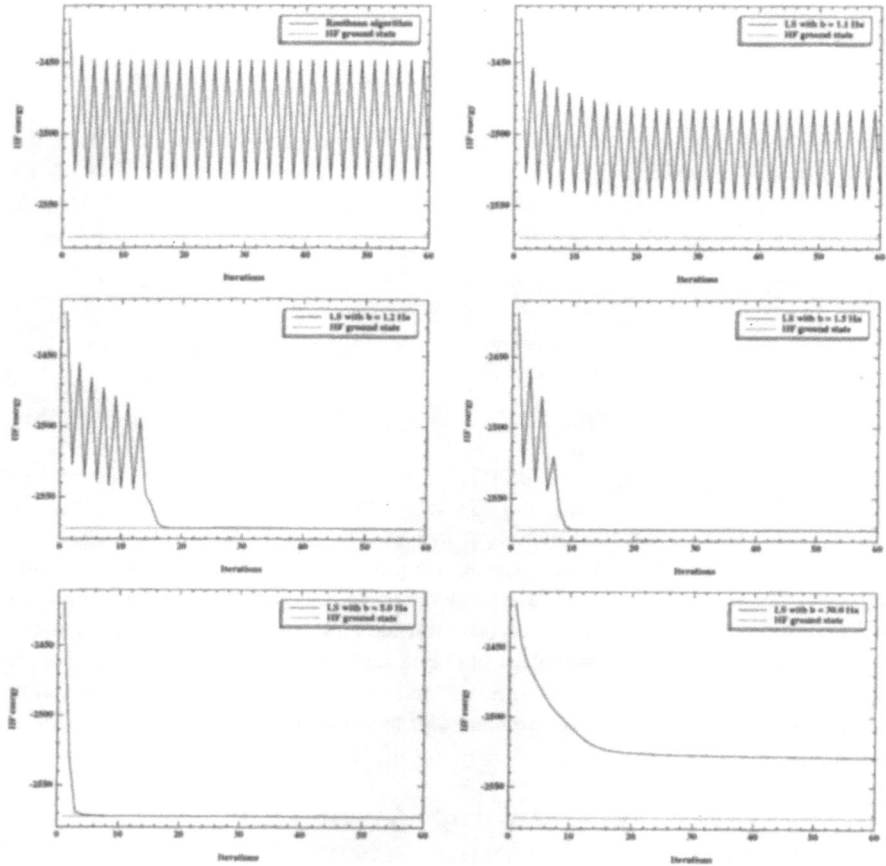

Figure 2.3: Calculation of the UHF ground state of the Bromine atom.

- The principal sequence of density matrices $(D_k)_{k \in \mathbf{N}}$ which will be proved to numerically converge towards an *aufbau* solution to the HF equations;

- A secondary sequence $(\widetilde{D}_k)_{k \geq 1}$ of pseudo-density matrices which belong to the set

$$\widetilde{\mathcal{P}} = \left\{ \widetilde{D} \in \mathcal{M}(n,n), \quad \widetilde{D}^* = \widetilde{D}, \quad \mathrm{Tr}\,(\widetilde{D}) = N, \quad \widetilde{D}^2 \leq \widetilde{D} \right\}$$

obtained from $\mathcal{P}$ by relaxing the nonlinear constraints $D^2 = D$.

The latter statement is a direct consequence of

Lemma 3. *The set $\mathcal{P}$ is convex,*

whose proof is postponed until the end of the present section. Indeed, $\widetilde{D}_0 = D_0 \in \mathcal{P} \subset \widetilde{\mathcal{P}}$ and, by induction, if $\widetilde{D}_k \in \widetilde{\mathcal{P}}$ then by convexity $\widetilde{D}_{k+1} \in \mathrm{Seg}[\widetilde{D}_k, D_{k+1}] \subset \widetilde{\mathcal{P}}$ since $D_{k+1} \in \mathcal{P} \subset \widetilde{\mathcal{P}}$.

The properties of the ODA are put together in the following theorem.

Theorem 3. *For any initial guess D_0 for which the ODA is UWP,*

1. *The sequence $E(\widetilde{D}_k)$ decreases towards a stationary value of the HF energy.*

2. *The sequence $(D_k)_{k \in \mathbb{N}}$ converges towards an aufbau solution to the HF equations.*

As a first step towards the understanding of the ODA, let us consider $\widetilde{D}_k \in \widetilde{\mathcal{P}}$ and $D' \in \mathcal{P}$, and let us compute the variation of the HF energy on the line segment

$$\mathrm{Seg}[\widetilde{D}_k, D'] = \left\{ (1 - \lambda)\widetilde{D}_k + \lambda D', \quad \lambda \in [0, 1] \right\}.$$

We obtain for any $\lambda \in [0, 1]$,

$$E^{HF}((1 - \lambda)\widetilde{D}_k + \lambda D')$$
$$= E^{HF}(\widetilde{D}_k) + \lambda \mathrm{Tr}\left(F(\widetilde{D}_k) \cdot (D' - \widetilde{D}_k) \right) + \frac{\lambda^2}{2} \mathrm{Tr}\left(G(D' - \widetilde{D}_k) \cdot (D' - \widetilde{D}_k) \right).$$

The "steepest descent" direction, i.e. the density matrix D for which the slope $s_{\widetilde{D}_k \to D} = \mathrm{Tr}\left(F(\widetilde{D}_k) \cdot (D - \widetilde{D}_k) \right)$ is minimum, is given by the solution to the minimization problem

$$D = \arg \inf \left\{ \mathrm{Tr}\left(F(\widetilde{D}_k) \cdot (D' - \widetilde{D}_k) \right), \quad D' \in \mathcal{P} \right\},$$

which also reads

$$D = \arg \inf \left\{ \mathrm{Tr}\left(F(\widetilde{D}_k) \cdot D' \right), \quad D' \in \mathcal{P} \right\}.$$

This is precisely the direction D_{k+1} obtained by the *aufbau* principle. The ODA can therefore be interpreted as a steepest descent algorithm in $\widetilde{\mathcal{P}}$. The practical implementation of the ODA is detailed in [4] for the RHF setting. The cost of one ODA iteration is approximatively the same as the cost of one iteration of the Roothaan algorithm (see [4] for details).

Figure 2.4 reports on a comparison between the ODA and the DIIS approaches for the calculation of the RHF ground state of the E form of n-methyl-2-nitrovinylamine (CH_3-NH-CH=CH-NO_2) in the basis 6-31G(d) (see [7]). The speed of convergence is estimated by computing the logarithm of the difference between the HF energy of the current density matrix and the (presumed) HF ground state energy. Calculations have been performed within Gaussian 98 [8]. The graph on the left hand side corresponds to an initial guess computed by a semi-empirical method. In this case, both algorithms converge but the speed of convergence of the DIIS algorithm is higher. From a general viewpoint, numerical tests performed until now demonstrate that the ODA is efficient for performing the early iterations of the SCF procedure; when the sequence (D_k) has reached a neighbourhood of a critical point of the HF problem, convergence can be accelerated either by resorting to iterative subspace techniques or by switching to a quadratically convergent algorithm [4]. On the

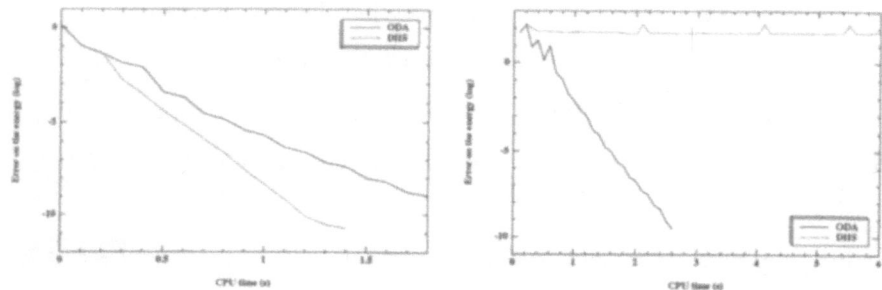

Figure 2.4: A comparison between the ODA and the DIIS algorithms: search for the RHF ground state of the E form of n-methyl-2-nitrovinylamine with an initial guess obtained by a semi-empirical method (on the left hand side) and with the initial guess obtained by diagonalizing the core hamiltonian (on the right hand side).

other hand, only the ODA converges for a more crude initial guess obtained by diagonalizing the core hamiltonian, as illustrated by the graph on the right hand side.

Let us now detail the

Proof of Theorem 3. Let us denote by $\tilde{F}_k = F(\widetilde{D}_k)$ and by $s_{k+1} = \mathrm{Tr}\,(\tilde{F}_k(D_{k+1} - \widetilde{D}_k))$. In view of Lemma 2,

$$s_{k+1} = \mathrm{Tr}\,\left(\tilde{F}_k D_{k+1}\right) - \mathrm{Tr}\,\left(\tilde{F}_k \widetilde{D}_k\right) \leq -\frac{\gamma}{2}\|D_{k+1} - \widetilde{D}_k\|^2.$$

As above, let us denote by b_0 a positive constant such that

$$\forall (\widetilde{D}, \widetilde{D}') \in \mathcal{M}(n,n) \times \mathcal{M}(n,n), \qquad \mathrm{Tr}\,\left(G(\widetilde{D}' - \widetilde{D}) \cdot (\widetilde{D}' - \widetilde{D})\right) \leq b_0 \|\widetilde{D} - \widetilde{D}'\|^2.$$

For any $\lambda \in [0,1]$,

$$E^{HF}((1-\lambda)\widetilde{D}_k + \lambda D_{k+1}) \leq E^{HF}(\widetilde{D}_k) - \frac{\gamma}{2}\|D_{k+1} - \widetilde{D}_k\|^2 \lambda + \frac{b_0}{2}\|D_{k+1} - \widetilde{D}_k\|^2 \lambda^2.$$

Therefore

$$\begin{aligned}
E^{HF}(\widetilde{D}_{k+1}) \\
= \inf\left\{E^{HF}((1-\lambda)\widetilde{D}_k + \lambda D_{k+1}), \qquad \lambda \in [0,1]\right\} \\
\leq \inf\left\{E^{HF}(\widetilde{D}_k) - \frac{\gamma}{2}\|D_{k+1} - \widetilde{D}_k\|^2\lambda + \frac{b_0}{2}\|D_{k+1} - \widetilde{D}_k\|^2\lambda^2, \qquad \lambda \in [0,1]\right\} \\
= E^{HF}(\widetilde{D}_k) - \alpha\|D_{k+1} - \widetilde{D}_k\|^2
\end{aligned}$$

with $\alpha = \gamma^2/8b_0$ if $\gamma \leq 2b_0$, $\alpha = (\gamma - b_0)/2$ otherwise. We then add up the above inequalities for $k \in \mathbb{N}$, and we get $\sum \|D_{k+1} - \widetilde{D}_k\|^2 < +\infty$, which implies that

$$D_{k+1} - \widetilde{D}_k \longrightarrow 0. \tag{10}$$

As $\widetilde{D}_{k+1} \in [\widetilde{D}_k, D_{k+1}]$, it follows that

$$\widetilde{D}_{k+1} - \widetilde{D}_k \longrightarrow 0,$$

and then that

$$D_{k+1} - D_k \longrightarrow 0.$$

Besides

$$\mathrm{Tr}\,(F(\widetilde{D}_k)D_{k+1}) = \inf\left\{\mathrm{Tr}\,(F(\widetilde{D}_k)D),\quad D \in \mathcal{P}\right\}. \tag{11}$$

Putting together (10) and (11), we finally obtain

$$\mathrm{Tr}\,(F(D_{k+1})D_{k+1}) - \inf\left\{\mathrm{Tr}\,(F(D_{k+1})D),\quad D \in \mathcal{P}\right\} \longrightarrow 0. \quad \Diamond$$

The following two points discuss the links between RCA and other algorithms like the level-shifting and the DIIS algorithms.

The first point concern the level-shifting algorithm. Let us use the ODA to minimize the penalized energy functional

$$E^b(D) = E^{HF}(D) - \frac{b}{2}\mathrm{Tr}\,(D^2).$$

As for any $D \in \mathcal{P}$, $\mathrm{Tr}\,(D^2) = \mathrm{Tr}\,(D) = N$, the critical points of the minimization problem

$$\inf\left\{E^b(D),\quad D \in \mathcal{P}\right\} \tag{12}$$

are the same as those of the HF problem (5). On the other hand, for any $\widetilde{D} \in \widetilde{\mathcal{P}} \setminus \mathcal{P}$, $\mathrm{Tr}\,(D^2) < N$: "interior" points are penalized. Let us denote by (D_k^b) and $(\widetilde{D}_k^b)$ the sequences generated by the ODA algorithm applied to (12):

1. Diagonalize the current Fock matrix $\tilde{F}_k^b = F(\widetilde{D}_k^b) - b\widetilde{D}_k^b$ and assemble the matrix $D_{k+1}^b \in \mathcal{P}_N$ by the *aufbau* principle;

2. Set $\widetilde{D}_{k+1}^b = \arg\inf\left\{E(\widetilde{D}),\quad \widetilde{D} \in \mathrm{Seg}[\widetilde{D}_k^b, D_{k+1}^b]\right\}$.

As for any $\lambda \in [0, 1]$

$$\begin{aligned}
&E^b((1-\lambda)\widetilde{D}_k^b + \lambda D_{k+1}^b) \\
={}& E^b(\widetilde{D}_k^b) + \lambda\mathrm{Tr}\,(\tilde{F}_k^b \cdot (D_{k+1} - \widetilde{D}_k)) \\
&+ \frac{\lambda^2}{2}\left(\mathrm{Tr}\,\left(G(D_{k+1}^b - \widetilde{D}_k^b) \cdot (D_{k+1}^b - \widetilde{D}_k^b)\right) - b\|D_{k+1}^b - \widetilde{D}_k^b\|^2\right),
\end{aligned}$$

we obtain

$$s_{k+1} = \mathrm{Tr}\,(\tilde{F}_k^b \cdot (D_{k+1} - \widetilde{D}_k)) \le -\frac{\gamma}{2}\|D_{k+1}^b - \widetilde{D}_k^b\|^2$$

and for $b \ge b_0$

$$\mathrm{Tr}\,\left(G(D_{k+1}^b - \widetilde{D}_k^b) \cdot (D_{k+1}^b - \widetilde{D}_k^b)\right) - b\|D_{k+1}^b - \widetilde{D}_k^b\|^2 \le 0.$$

For $b \geq b_0$, the function $\lambda \mapsto E^b((1 - \lambda)\widetilde{D}_k^b + \lambda D_{k+1}^b)$ is therefore decreasing and concave on $[0, 1]$; it follows that $\widetilde{D}_{k+1}^b = D_{k+1}^b$ which means that the ODA for minimizing (12) coincides with the level-shifting algorithm. This provides in particular another proof of the convergence of the level-shifting algorithm for large shift parameters b. Now, if D^b is an accumulation point of the sequence (D_k^b), we obtain by passing to the limit

$$s_\infty = \inf\left\{ \mathrm{Tr}\left((F(D^b) - bD^b) \cdot (D - D^b)\right), \quad D \in \mathcal{P} \right\} = 0.$$

This implies that

$$D^b = \arg\inf\left\{ \mathrm{Tr}\left((F(D^b) - bD^b) \cdot D\right), \quad D \in \mathcal{P} \right\},$$

and therefore that

$$[F(D^b) - bD^b, D^b] = [F(D^b), D^b] = 0,$$

but not necessarily that $D^b = \arg\inf\left\{ \mathrm{Tr}\left(F(D^b)D\right), \quad D \in \mathcal{P} \right\}$: D^b is a solution to the HF equations that may not satisfy the *aufbau* principle. In addition, as $G(D) \geq 0$ for any $D \in \mathcal{P}$, we obtain

$$
\begin{aligned}
s_{k+1} &= \mathrm{Tr}\left(\widetilde{F}_k^b \cdot (D_{k+1}^b - D_k^b)\right) \\
&= \mathrm{Tr}\left(F(D_k^b) \cdot (D_{k+1}^b - D_k^b)\right) + \frac{b}{2}\|D_{k+1}^b - D_k^b\|^2 \\
&\geq -2\left|\inf\left\{ \mathrm{Tr}\,(hD), \quad D \in \mathcal{P} \right\}\right| + \frac{b}{2}\|D_{k+1}^b - D_k^b\|^2.
\end{aligned}
$$

It results that

$$\|D_{k+1}^b - D_k^b\|^2 \leq \frac{2}{b}\left|\inf\left\{ \mathrm{Tr}\,(hD), \quad D \in \mathcal{P} \right\}\right|.$$

The level-shifting algorithm can then also be interpreted as a trust region algorithm on the manifold $\mathcal{P}$ for which the radius of the trust region is bounded by $\delta = \frac{2}{b}\left|\inf\left\{ \mathrm{Tr}\,(hD), \quad D \in \mathcal{P} \right\}\right|$. The larger the shift parameter b, the smaller the step $D_{k+1}^b - D_k^b$; this induces for large b a slow motion along a steepest descent path.

The second point is related to the DIIS algorithm. An attempt of improvement of the ODA consists, in the spirit of iterative subspace methods, in keeping in memory all (or some of) the density matrices computed at the previous steps and in minimizing the HF energy in the convex set generated by all the density matrices stored in memory:

1. Diagonalize $F(\widetilde{D}_k)$ and assemble the density matrix $D_{k+1} \in \mathcal{P}$ by the *aufbau* principle .

2. Set $\widetilde{D}_{k+1} = \arg\inf\left\{ E^{HF}(\widetilde{D}), \quad \widetilde{D} = \sum_{i=0}^{k+1} c_i D_i, \quad 0 \leq c_i \leq 1, \quad \sum_{i=0}^{k+1} c_i = 1 \right\}.$

This algorithm is similar to Pulay's DIIS algorithm [20] except that in the DIIS algorithm, step 2 consists in minimizing the residual

$$\left\|\sum_{i=0}^{k+1} c_i[F(D_i), D_i]\right\|^2, \tag{13}$$

where $[.,.]$ denotes the commutator $[A, B] = AB - BA$, and where $\| \cdot \|$ denotes the Hilbert-Schmidt norm. Contrary to the RCA presented here, the DIIS algorithm may diverge: the residual (13) actually decreases at each step but it may vanish without the convergence is met.

Let us conclude this section with the

Proof of Lemma 3. Let $\widetilde{D}_1 \in \mathcal{P}$ and $\widetilde{D}_2 \in \mathcal{P}$. For any $0 \leq c_1, c_2 \leq 1$ such that $c_1 + c_2 = 1$,

$$
\begin{aligned}
\widetilde{D}^2 &= (c_1 \widetilde{D}_1 + c_2 \widetilde{D}_2)^2 \\
&= c_1^2 \widetilde{D}_1^2 + c_2^2 \widetilde{D}_2^2 + c_1 c_2 (\widetilde{D}_1 \widetilde{D}_2 + \widetilde{D}_2 \widetilde{D}_1) \\
&= c_1 \widetilde{D}_1 + c_2 \widetilde{D}_2 + c_1 (\widetilde{D}_1^2 - \widetilde{D}_1) + c_2 (\widetilde{D}_2^2 - \widetilde{D}_2) \\
&\quad + (c_1^2 - c_1) \widetilde{D}_1^2 + (c_2^2 - c_2) \widetilde{D}_2^2 + c_1 c_2 (\widetilde{D}_1 \widetilde{D}_2 + \widetilde{D}_2 \widetilde{D}_1) \\
&= \widetilde{D} + c_1 (\widetilde{D}_1^2 - \widetilde{D}_1) + c_2 (\widetilde{D}_2^2 - \widetilde{D}_2) - c_1 c_2 (\widetilde{D}_1 - \widetilde{D}_2)^2
\end{aligned}
$$

since $c_1^2 - c_1 = -c_1(1 - c_1) = -c_1 c_2 = c_2^2 - c_2$. Now, $\widetilde{D}_1^2 - \widetilde{D}_1 \leq 0$, $\widetilde{D}_2^2 - \widetilde{D}_2 \leq 0$ and $(\widetilde{D}_1 - \widetilde{D}_2)^2 \geq 0$. Consequently, $\widetilde{D}^2 \leq \widetilde{D}$. The other two constraints ($\widetilde{D} = \widetilde{D}^*$ and $\text{Tr}\,(\widetilde{D}) = N$) being linear, it is clear that $\widetilde{D} \in \widetilde{\mathcal{P}}$. $\Diamond$

Bibliography

[1] V. Bach, E.H. Lieb, M. Loss and J.P. Solovej, *There are no unfilled shells in unrestricted Hartree-Fock theory*, Phys. Rev. Letters 72 (1994) 2981-2983.

[2] V. Bonačić-Koutecký and J. Koutecký, *General properties of the Hartree-Fock problem demonstrated on the frontier orbital model. II. Analysis of the customary iterative procedure*, Theoret. Chim. Acta 36 (1975) 163-180.

[3] E. Cancès and C. Le Bris, *On the convergence of SCF algorithms for the Hartree-Fock equations*, to appear in Math. Model. Num. Anal.

[4] E. Cancès and C. Le Bris, *Can we outperform the DIIS approach for electronic structure calculations*, to appear in Int. J. Quantum Chem.

[5] J.-M. Combes, P. Duclos and R. Seiler, *The Born-Oppenheimer approximation*, in *Rigorous atomic and molecular physics*, G. Velo and A. Wightman (Eds), Plenum Press 1981.

[6] R. Fletcher, *Optimization of SCF LCAO wave functions*, Mol. Phys. 19 (1970) 55-63.

[7] J.B. Foresman and A. Frisch, *Exploring chemistry with electronic structure methods*, 2nd edition, Gaussian Inc., Pittsburgh PA 1996.

[8] M.J. Frisch, G.W. Trucks, H.B. Schlegel, G.E. Scuseria, M.A. Robb, J.R. Cheeseman, V.G. Zakrzewski, J.A. Montgomery, R.E. Stratmann, J.C. Burant, S. Dapprich, J.M. Millam, A.D. Daniels, K.N. Kudin, M.C. Strain, O. Farkas, J. Tomasi, V. Barone, M. Cossi, R. Cammi, B. Mennucci, C. Pomelli, C. Adamo, S. Clifford, J. Ochterski, G.A. Petersson, P.Y. Ayala, Q. Cui, K. Morokuma, D.K. Malick, A.D. Rabuck, K. Raghavachari, J.B. Foresman, J. Cioslowski, J.V. Ortiz, B.B. Stefanov, G. liu, A. Liashenko, P. Piskorz, I. Kpmaromi, G. Gomperts, R.L. Martin, D.J. Fox, T. Keith, M.A. Al-Laham, C.Y. Peng, A. Nanayakkara, C. Gonzalez, M. Challacombe, P.M.W. Gill, B.G. Jpohnson, W. Chen, M.W. Wong, J.L. Andres, M. Head-Gordon, E.S. Replogle and J.A. Pople, Gaussian 98 (Revision A.7), Gaussian Inc., Pittsburgh PA 1998.

[9] A. Frisch and M.J. Frisch, *Gaussian 98 user's reference*, Gaussian Inc., Pittsburgh PA 1999.

[10] D.R. Hartree, *The calculation of atomic structures*, Wiley 1957.

[11] A. Igawa and H. Fukutome, *A new direct minimization algorithm for Hartree-Fock calculations*, Prog. Theor. Phys. 54 (1975) 1266-1281.

[12] J. Koutecký and V. Bonačić, *On convergence difficulties in the iterative Hartree-Fock procedure*, J. Chem. Phys. 55 (1971) 2408-2413.

[13] E.H. Lieb, *Bound on the maximum negative ionization of atoms and molecules*, Phys. Rev. A 29 (1984) 3018-3028.

[14] E.H. Lieb and B. Simon, *The Hartree-Fock theory for Coulomb systems*, Commun. Math. Phys. 53 (1977) 185-194.

[15] P.-L. Lions, *Solutions of Hartree-Fock equations for Coulomb systems*, Comm. Math. Phys. 109 (1987) 33-97.

[16] P.-L. Lions, *Hartree-Fock and related equations*, Nonlinear partial differential equations and their applications, Collège de France Seminar Vol. 9 (1988) 304-333.

[17] R. McWeeny, *The density matrix in self-consistent field theory I. Iterative construction of the density matrix*, Proc. R. Soc. London Ser. A 235 (1956) 496-509.

[18] R. McWeeny, *Methods of molecular Quantum Mechanics*, Academic Press 1992.

[19] A. Neumaier, *Molecular modeling of proteins and mathematical prediction of protein structure*, SIAM Rev. 39 (1997) 407-460.

[20] P. Pulay, *Improved SCF convergence acceleration*, J. Comp. Chem. 3 (1982) 556-560.

[21] C.C.J. Roothaan, *New developments in molecular orbital theory*, Rev. Mod. Phys. 23 (1951) 69-89.

[22] V.R. Saunders and I.H. Hillier, *A "level-shifting" method for converging closed shell Hartree-Fock wave functions*, Int. J. Quantum Chem. 7 (1973) 699-705.

[23] H.B. Schlegel and J.J.W. McDouall, *Do you have SCF stability and convergence problems?*, in *Computational Advances in Organic Chemistry*, Kluwer Academic, 1991, 167-185.

[24] T. Schlick, *Optimization methods in computational chemistry*, in *Reviews in Computational Chemistry, Vol. III*, K.B. Lipkowitz and D.B. Boyd (Eds.), VCH Publishers 1992.

[25] R. Seeger R. and J.A. Pople, *Self-consistent molecular orbital methods. XVI. Numerically stable direct energy minimization procedures for solution of Hartree-Fock equations*, J. Chem. Phys. 65 (1976) 265-271.

[26] R.E. Stanton, *The existence and cure of intrinsic divergence in closed shell SCF calculations*, J. Chem. Phys. 75 (1981) 3426-3432.

[27] R.E. Stanton, *Intrinsic convergence in closed-shell SCF calculations. A general criterion*, J. Chem. Phys. 75 (1981) 5416-5422.

[28] M.C. Zerner and M. Hehenberger, *A dynamical damping scheme for converging molecular SCF calculations*, Chem. Phys. Letters 62 (1979) 550-554.

Chapter 3

A pedagogical introduction to Quantum Monte-Carlo

Michel Caffarel [1] & **Roland Assaraf** [2]

[1] *CNRS-Laboratoire de Chimie Théorique,*
Tour 22-23, Université Pierre et Marie Curie,
4 place Jussieu 75252 Paris Cedex 05 France.
mc@lct.jussieu.fr
[2] *Scuola Internazionale Superiore di Studi Avanzati (SISSA),*
and Istituto Nazionale Fisica della Materia (INFM),
Via Beirut 4, I-32013 Trieste, Italy
assaraf@sissa.it

Abstract : Quantum Monte Carlo (QMC) methods are powerful stochastic approaches to calculate ground-state properties of quantum systems. They have been applied with success to a great variety of problems described by a Schrödinger-like Hamiltonian (quantum liquids and solids, spin systems, nuclear matter, *ab initio* quantum chemistry, etc...). In this paper we give a pedagogical presentation of the main ideas of QMC. We develop and exemplify the various concepts on the simplest system treatable by QMC, namely a 2×2 matrix. First, we discuss the Pure Diffusion Monte Carlo (PDMC) method which we consider to be the most natural implementation of QMC concepts. Then, we discuss the Diffusion Monte Carlo (DMC) algorithms based on a branching (birth-death) process. Next, we present the Stochastic Reconfiguration Monte Carlo (SRMC) method which combines the advantages of both PDMC and DMC approaches. A very recently introduced optimal version of SRMC is also discussed. Finally, two methods for accelerating QMC calculations are sketched: (a) the use of integrated Poisson processes to speed up the dynamics (b) the introduction of "renormalized" or "improved" estimators to decrease the statistical fluctuations.

1 Introduction

Quantum Monte Carlo methods are powerful methods to compute the ground-state properties of quantum systems. They have been applied with success to a large variety of problems described by a Schrödinger-like equation (see, *e.g.*, [1],[2],[3], [4],[5]). Although many variants can be found in the literature all the methods rest on the same idea. In essence, QMC approaches are power methods [6]. Starting from some approximation of the ground-state, some simple function of the Hamiltonian matrix (or operator for continuous spaces) is applied a large number of times on it and the convergence to the exact ground-state is searched for. In contrast with the usual power method the multiplication -matrix times vector- is not done exactly by summing explicitly all components of the current vector but estimated in average through some stochastic sampling. Such a procedure becomes particularly interesting when the linear space has a size prohibiting any attempt of multiplying or even storing matrices (say, a size much larger than 10^9). For such problems QMC is in general the most reliable and accurate method available. The price to pay is that the stochastic sampling of the product -matrix times vector- at the heart of power methods has to be done in a very efficient way. The key ingredient to realize this is to introduce some "importance sampling" techniques. This is done via a trial vector whose role is to guide the stochastic "walkers" (points of the stochastic chain built on the computer) in the "important" regions (regions corresponding to an important contribution to the averages). In addition to this, it can be shown that there exists a most important "zero-variance property": the closer the trial vector is to the exact one, the smaller the statistical fluctuations are, and the more rapid the convergence to the exact eigenvector is. Accordingly, the efficiency of QMC methods is directly related to the quality of the trial vector used. It should be noted that powerful methods have been developed to construct and optimize trial vectors within a QMC framework [7].

A second central aspect determining the efficiency of QMC is the way the sign of the exact wave function is sampled. For distinguishable particles or bosons described by a standard Hamiltonian operator (kinetic + potential terms) it can be shown that the ground-state components (defined in the usual basis) have a constant sign (say, positive). Therefore, they can be directly interpreted as a probabilistic density and no "sign problem" arises. Highly accurate simulations of systems including thousands of bosons have been done [2] and QMC methods can be considered as the most reliable and powerful approaches for computing the ground-state properties of such systems. In contrast, for fermions the wave function has no longer a constant sign (as a consequence of the Pauli principle). We are then faced to the problem of sampling a quantity -the sign- which fluctuates a lot (regions of configuration space with different signs are highly intricate [8]) and has essentially zero-average (because of the orthogonality between the Fermi and Bose ground-states). Even worse, it can be shown that the signed (fermionic) stochastic averages which are expressed as a difference of two positive (bosonic) quantities (one associated with the positive part of the wave function, the other with the negative one) have a statistical error which increases *exponentially* with the number of iterations of the power method [9],[10],

[11] (with a constant in the exponential which also increases with the number of fermions!). This fundamental problem is known as the "sign problem". Its solution is one of the central issues of present-day computational physics. Despite this major difficulty, to simulate fermions is not a desperate task. A first commonly employed solution consists in removing the sign from the simulations by imposing to the walkers to remain in regions of constant sign for the trial function. This approach, known as the fixed-node method, is stable (we define a set of independent sign-free bosonic simulations in each nodal domain) but is not exact due to the approximate location of the trial nodes (fixed-node bias). The quality of the fixed-node approximation is very dependent on the specific system studied and on the nature of the physics involved. In some cases the result obtained can be very satisfactory [12] (at least for the total energies). On the other hand, a number of exact schemes for treating fermions have also been develozped such as transient methods, nodal release approaches, etc... In short, it is found that exact fermion simulations can be made to converge in a reasonable amount of CPU time only when the difference between the Fermi and Bose ground-state energies is not too large and/or when the starting trial wave function is good enough.

We conclude these general considerations by some remarks about the application of QMC methods to *ab initio* quantum chemistry. The calculation of electronic properties of atoms and molecules is certainly one of the most difficult applications for QMC techniques. The reasons are two-fold. First, as already emphasized above, electrons are fermions and the sign of the wave function has to be properly taken into account. However, as a number of actual simulations have shown, this problem is not dramatic for molecules (at least for the calculation of total electronic energies). By using the nodes of Hartree-Fock wave functions accurate energies can be obtained [12]. Second, and this is the most severe difficulty, the electronic density resulting from the Pauli principle (shell-structure) and the highly inhomogeneous nuclear coulombic attraction (strongly attractive nuclear centers at localized positions) displays a very intricate and subtle structure. In particular, the simulation must properly take into account the very different energy scales present in atoms (both high-energy core electrons and low-energy valence electrons). Since we are mainly interested in small differences of energies (bond energies, affinities, etc...) any error in the regions where the density of energy is large can be responsible for uncontrolled fluctuations in low-energy regions of physical or chemical interest. As a consequence, highly accurate trial wave functions are needed to get reliable results[13],[14]. Using as trial wave functions standard approximations of quantum chemistry (Hartree-Fock, small CI, etc...) and adding to them some correlated part including explicitly the interelectronic coordinate, a number of all-electron calculations of total energies have been performed for small atoms and molecules[4],[5]. Results are excellent (essentially, the exact values are obtained). The maximum number of electrons tractable is very dependent on the number of nuclei, the magnitude of the nuclear charges, the quality of the trial wave function, etc... However, for more than a dozen of electrons all-electron simulations can rapidly become very time consuming. For more electrons all-electron calculations are almost impossible because of the large statistical fluctuations resulting from the core electrons. A natural solution consists in freezing in some way the high-energy core electrons as it is routinely done in *ab initio* quantum chemistry with basis sets. Without entering into the details of the various proposals presented in the QMC literature (model potentials [15], pseudohamiltonians[16], non-local pseudopotentials [17],[18], etc...) it is important to mention that some simple and stable schemes are now at our disposal. By using pseudopotentials and the fixed-node approximation it is possible to restrict the simulations to valence electrons only in a coherent way [19],[20], [21]. This represents a real breakthrough toward a systematic use of QMC techniques in *ab initio* quantum chemistry. In order to illustrate what QMC can achieve we present in Table 3.1 a short and very arbitrary selection of results obtained for total energies and binding energies of several atoms and molecules. Of course, much more applications have been done (in particular calculations of properties other than energy, excited states, etc...) the interested readers are referred to the literature (for example, [4],[5]).

Table 3.1: A short selection of QMC calculations for atoms and molecules. Total energies results correspond to all-electron fixed-node (approximate) or release-node (exact) calculations. Results for the binding energies are fixed-node calculations with pseudopotentials.

Total energies (atomic units)	E_0(Hartree-Fock)	E_0(QMC)	E_0("exact"[a])
He[no approx.]	-2.86168^b	$-2.9037244(1)^c$	-2.90372437^d
H_2[no approx.]	-1.13363	$-1.174(1)^e$	-1.17447^f
LiH[fixed-node]	-7.987	$-8.0680(6)^g$	-8.07021^a
LiH[release-node]	-7.987	$-8.07021(5)^h$	-8.07021^a
H_2O[fixed-node]	-76.0675	$-76.377(7)^i$	-76.4376^a
F_2[fixed-node]	-198.7701	$-199.487(1)^j$	-199.5299
Binding energies(eV) of hydrocarbons[k]	HF[l]	LDA[m]	QMC	Exp.
Methane (CH_4)	14.20	20.59	18.28(5)	18.19
Acytylene (C_2H_2)	12.70	20.49	17.53(5)	17.59
Ethane (C_2H_6)	23.87	35.37	31.10(5)	30.85
Benzene (C_6H_6)	44.44	70.01	59.2(1)	59.24
Binding energies(eV/atom) of Si clusters[n]	HF[l]	LDA[m]	QMC	Exp.
$Si_2(D_{2h})$	0.85	1.98	1.580(7)	1.61(4)
$Si_3(C_{3v})$	1.12	2.92	2.374(8)	2.45(6)
$Si_7(D_{5h})$	1.91	4.14	3.43(2)	3.60(4)
$Si_{10}(C_{3v})$	1.89	4.32	3.48(2)	...
$Si_{20}(C_{3v})$	1.55	4.28	3.43(3)	...

[a] From experimental data analysis; [b] Ref.[22]; [c] Ref.[14]; [d] Ref.[23]; [e] Ref.[24]; [f] Ref.[25]; [g] Ref.[30]; [h] Ref.[31]; [i] Ref.[10]; [j] Ref.[32]; [k] From Ref.[33]; [l] Hartree-Fock approximation; [m] Local Density Approximation (Density Functional Theory); [n] From Ref.[34];

In this paper we attempt to give a pedagogical presentation of QMC. It is intended to newcomers in the field or any person interested in the general ideas of

QMC. As we shall see, QMC is a method of very broad applicability. It can be used for studying problems defined in continuous configuration space (Schrödinger equation for particles moving in a continuum), for models described by an infinite matrix (*e.g.* problems defined on a lattice), and of course for finite matrices. In order to focus on QMC and not on problems arising from particular physical systems, we develop and exemplify the various concepts on the simplest system treatable by QMC, namely a 2×2 matrix. We have chosen to use the standard mathematical notation employed in linear algebra instead of the most specialized notation employed in quantum physics (use of Dirac's notation in terms of "bras" and "kets").

The contents of the paper is as follows. In Sec. 2.1. we introduce our toy problem (matrix 2×2). In Sec. 2.2 the central idea of QMC approaches is presented: In essence, QMC can be viewed as a stochastic implementation of the power method. Section 2.3 is devoted to the presentation of the fundamental relation between the matrix to be diagonalized and a stochastic matrix. The simplest expression for the energy resulting from this connection is then given. Sec. 2.4 presents the Pure Diffusion Monte Carlo (PDMC) method which we consider to be the most natural implementation of QMC concepts. Then, in Sec. 2.5 we discuss the Diffusion Monte Carlo (DMC) algorithms based on a branching (birth-death) process. Next, we present in Sec. 2.6 the Stochastic Reconfiguration Monte Carlo (SRMC) method which combines the advantages of both PDMC and DMC approaches. An optimal version of SRMC very recently introduced is also discussed. In Sec. 2.7 two methods for accelerating QMC calculations are sketched: (a) the use of integrated Poisson processes to speed up the dynamics (Sec. 2.7.1) (b) the introduction of "renormalized" or "improved" estimators to decrease the statistical fluctuations (Sec. 2.7.2). Finally, in Sec. 3 we present a number of aspects related to the implementation of QMC to an arbitrary problem (QMC for a $N \times N$ matrix: Sec. 3.1; QMC in a continuum: Sec. 3.2).

2   QMC for a 2×2 matrix

2.1   The 2×2 matrix

The simplest system treatable by QMC is a 2×2 matrix. Let us denote by H this matrix. It is written under the form

$$H = \begin{pmatrix} v_0 & -t \\ -t & v_1 \end{pmatrix} \tag{1}$$

where, without loss of generality, v_0 is taken smaller or equal to v_1 (the labels for the two basis vectors can be chosen arbitrarily) and t is chosen positive (choice of the overall sign for the matrix). The lowest eigenvalue (ground-state energy) is given by

$$E_0 = \frac{v_0 + v_1 - \sqrt{(v_1 - v_0)^2 + 4t^2}}{2} \tag{2}$$

and the corresponding ground-state eigenvector is

$$\mathbf{u_0} = \begin{pmatrix} \frac{t}{\sqrt{(E_0-v_0)^2+t^2}} \\ \frac{v_0-E_0}{\sqrt{(E_0-v_0)^2+t^2}} \end{pmatrix} \tag{3}$$

Note that $E_0 \leq v_0$ and that both components of $\mathbf{u_0}$ are positive.

2.2   QMC is a power method

The central idea of QMC is to extract from a known trial vector $\mathbf{v}$ its exact ground-state component, $\mathbf{u_0}$. This is realized by using a matrix $G(H)$ acting as a filter:

$$\lim_{n\to\infty} G(H)^n\mathbf{v} \sim \mathbf{u_0} \tag{4}$$

where H is the original matrix. This idea is similar to what is done in power methods where the maximal eigenvector (associated with the eigenvalue of maximum modulus) is obtained by applying a large number of times the matrix on an arbitrary initial vector [6]. A most popular choice for $G(H)$ is

$$G(H) \equiv 1 - \tau(H - E_T) \tag{5}$$

where "1" denotes the identity matrix ($[1]_{ij} = \delta_{ij}$) and τ is some positive constant (a "time-step" in physical applications) to be adjusted (see below) and E_T is some arbitrary reference energy.

In our toy example, the initial trial vector $\mathbf{v}$ can be decomposed within the basis defined by the eigenvectors of H:

$$\mathbf{v} = c_0\mathbf{u_0} + c_1\mathbf{u_1}. \tag{6}$$

The action of $G(H)$ at iteration number n defined as follows

$$\mathbf{v}^{(n)} \equiv G(H)^n\mathbf{v} = [1 - \tau(H - E_T)]^n\mathbf{v} \tag{7}$$

can therefore be expressed as

$$\mathbf{v}^{(n)} = c_0[1 - \tau(E_0 - E_T)]^n\mathbf{u_0} + c_1[1 - \tau(E_1 - E_T)]^n\mathbf{u_1}. \tag{8}$$

Provided that c_0 is non-zero, that is, there exists some overlap between the trial vector $\mathbf{v}$ and the exact solution, the vector $\mathbf{v}^{(n)}$ will converge geometrically to the exact ground-state eigenvector (up to some inessential normalization factor). The correction at finite n is essentially exponential in n and in the gap in energy $E_1 - E_0$

$$\mathbf{v}^{(n)} \sim_{n\to\infty} \mathbf{u_0} + O[\exp(-n\tau(E_1 - E_0))]. \tag{9}$$

Note that this result is independent on the value of E_T if τ is chosen positive. In practice, the exact eigenvalue can be extracted by considering the large-n limit of the following ratio, known as the Rayleigh quotient

$$E_0(n) = \frac{\mathbf{v}^T H \mathbf{v}^{(n)}}{\mathbf{v}^T \mathbf{v}^{(n)}} \sim_{n\to\infty} E_0 \tag{10}$$

Finally, the problem reduces to compute the action of the matrix H over some arbitrary vector, $\mathbf{v}^{(n)}$. When the size of the linear space is small this can be easily done by a simple matrix multiplication. Here, we have

$$c_0^{(n+1)} = [1 - \tau(v_0 - E_T)]c_0^{(n)} + t\tau c_1^{(n)}$$

$$c_1^{(n+1)} = t\tau c_0^{(n)} + [1 - \tau(v_1 - E_T)]c_1^{(n)} \tag{11}$$

where $c_i^{(k)}$ are the two components of the vector $\mathbf{v}^{(k)}$.

In Table 3.2 the convergence of $c_0^{(n)}, c_1^{(n)}$, and $E_0(n)$ are shown. The parameters chosen are the following: $v_0 = 1, v_1 = 2, t = 1, \tau = 0.1, E_T = 3$, and $c_0^{(0)} = c_1^{(0)} = 1/\sqrt{2}$. As it is seen the convergence to the exact values is quite rapid as a function of the iteration number n.

Table 3.2: Convergence of $c_0^{(n)}, c_1^{(n)}$, and $E_0(n)$ as a function of n.

n	$c_0^{(n)}$	$c_1^{(n)}$	$E_0^{(n)}$
$n = 10$	0.83253...	0.55397...	0.403295...
$n = 20$	0.84813...	0.52978...	0.385023...
$n = 30$	0.85029...	0.52630...	0.382401...
$n = 40$	0.85060...	0.52581...	0.382027...
$n = 50$	0.85064...	0.52574...	0.381974...
$n = 60$	0.85064...	0.52573...	0.381967...
$n = 70$	0.85065...	0.52573...	0.381966...
$n = 80$	0.85065...	0.52573...	0.381966...
Exact values	0.85065...[a]	0.52573...[a]	0.381966...[b]

[a] See equation (3).
[b] See equation (2).

Note that the convergence as a function of the iteration number n of the power method can be greatly enhanced by using all information present in the set of iterated vectors, $\mathbf{v}^{(p)}, p = 0...n$. This idea is implemented by introducing a so-called Krylov space [26],[27] associated with the pair $(H, \mathbf{v})$ and defined as the linear space spanned by the vectors $\mathbf{v}^{(p)}, p = 0...n$. By diagonalizing the matrix H within this subspace the optimal solution at iteration n is obtained under the form $\mathbf{u} = \sum_{i=0}^{n} c_i \mathbf{v}^{(i)}$ instead of $\mathbf{u} = \mathbf{v}^{(n)}$, and the convergence as a function of n is much more rapid. Many variations of this idea have been presented in the literature for treating "big" matrices of size ranging from 10^3 to 10^9. Among them two well-known variants are the Lanczòs [26] and Davidson [28] algorithms. For matrices of much larger size, stochastic approaches are to be used.

2.3 Introducing a stochastic matrix

In order to perform the product -matrix times vector- stochastically we shall rewrite the matrix elements $[1 - \tau(H - E_T)]_{ij}$ in terms of a so-called stochastic matrix

$P_{i \to j}(\tau)$ and some residual contribution w_{ij} ("weight" matrix). As we shall see below, $P_{i \to j}(\tau)$ will be interpreted as the probability of moving from state i to state j and will be used to construct a random sequence of states. To be valid this probabilistic interpretation requires the two following conditions. First, to define a probability, the matrix elements must all be positive:

$$P_{i \to j}(\tau) \geq 0 \quad . \tag{12}$$

Second, the sum over final states must be equal to one:

$$\sum_j P_{i \to j}(\tau) = 1 \quad . \tag{13}$$

An arbitrary matrix is said to be stochastic when conditions (12) and (13) are fulfilled. Here, we define $P_{i \to j}(\tau)$ as follows

$$P_{i \to j}(\tau) = \frac{v_j}{v_i}[1 - \tau(H - E_L)]_{ij}. \tag{14}$$

where E_L is a diagonal matrix written as

$$[E_L]_{ij} = \delta_{ij} E_L(i) \tag{15}$$

and the vector $E_L(i)$ -called the local energy- is given by

$$E_L(i) = \frac{[H\mathbf{v}]_i}{v_i}. \tag{16}$$

Here τ is a parameter which will play the role of a time step. Note that the magnitude of the fluctuations of the local energy is a measure of the quality of the trial vector used. The closer the trial vector is to the exact one, the smaller the fluctuations are. In the optimal case corresponding to $\mathbf{v} = \mathbf{u_0}$, the local energy becomes constant (independent on the state i) and equal to the exact energy.

Now, using the definition of the stochastic matrix our fundamental quantity $[1 - \tau(H - E_T)]_{ij}$ can be rewritten as

$$[1 - \tau(H - E_T)]_{ij} = \frac{v_i}{v_j} P_{i \to j}(\tau) w_{ij} \tag{17}$$

where the weight w_{ij} is defined as

$$w_{ij} = \frac{[1 - \tau(H - E_T)]_{ij}}{[1 - \tau(H - E_L)]_{ij}}. \tag{18}$$

This quantity represents the residual contribution resulting from the fact that the trial vector is not exact. When $\mathbf{v} = \mathbf{u_0}$, all weights reduce to a constant (which can be made equal to one by taking $E_T = E_0$).

In our simple example the stochastic matrix writes

$$\begin{pmatrix} P_{0 \to 0}(\tau) & P_{1 \to 0}(\tau) \\ P_{0 \to 1}(\tau) & P_{1 \to 1}(\tau) \end{pmatrix} = \begin{pmatrix} 1 - \tau t \frac{v_1}{v_0} & \tau t \frac{v_0}{v_1} \\ \tau t \frac{v_1}{v_0} & 1 - \tau t \frac{v_0}{v_1} \end{pmatrix} \tag{19}$$

Since the quantities t, τ, and v_i are positive the diagonal elements can be made positive by choosing the time-step τ small enough:

$$\tau \leq \text{Min}[\frac{v_0}{v_1 t}, \frac{v_1}{v_0 t}] \tag{20}$$

Off-diagonal elements being also positive, condition (12) is fulfilled. We also check from (19) that $P_{0 \to 0}(\tau) + P_{0 \to 1}(\tau) = 1$ (starting from 0 the probability of arriving at 0 or 1 is equal to one) and $P_{1 \to 0}(\tau) + P_{1 \to 1}(\tau) = 1$ (*idem* when starting from 1), so that condition (13) is also valid. The general case is not difficult to verify. Condition (12) implies the following constraint on τ

$$0 \leq \tau \leq \frac{1}{\text{Max}_i[H_{ii} - E_L(i)]} \tag{21}$$

and condition (13) results from (14) and (16).

Let us now write the quantities of interest in terms of the stochastic matrix. Here, we shall focus our attention on the Rayleigh quotient (10) and, more fundamentally, on the expression of the n-th order iterated vector $\mathbf{v}^{(n)}$

$$\mathbf{v}^{(n)} = [1 - \tau(H - E_T)]^n \mathbf{v} \tag{22}$$

Writing explicitly the product of matrices present in (22), the components of the vector $\mathbf{v}^{(n)}$ can be written as

$$v_{i_n}^{(n)} = \sum_{i_{n-1}} \cdots \sum_{i_0} [1 - \tau(H - E_T)]_{i_n i_{n-1}} \cdots [1 - \tau(H - E_T)]_{i_1 i_0} v_{i_0} \tag{23}$$

Introducing the weight matrix (18) we obtain

$$v_{i_n}^{(n)} = \sum_{i_{n-1}} \cdots \sum_{i_0} w_{i_n i_{n-1}} [1 - \tau(H - E_L)]_{i_n i_{n-1}} \cdots w_{i_1 i_0} [1 - \tau(H - E_L)]_{i_1 i_0} v_{i_0} \tag{24}$$

and, finally (after permuting the indices and using the definition (14) of the stochastic matrix)

$$v_{i_n}^{(n)} = \sum_{i_0} \cdots \sum_{i_{n-1}} v_{i_0}^2 P_{i_0 \to i_1} \cdots P_{i_{n-1} \to i_n} w_{i_0 i_1} \cdots w_{i_{n-1} i_n} \frac{1}{v_{i_n}}. \tag{25}$$

Now, by defining the following quantity:

$$P(i_0, ..., i_n) = v_{i_0}^2 P_{i_0 \to i_1} \cdots P_{i_{n-1} \to i_n} \tag{26}$$

we can rewrite the energy at iteration n as follows:

$$E_0(n) = \frac{\sum_{i_0} \cdots \sum_{i_n} P(i_0, ..., i_n) E_L(i_n) \prod_{k=0}^{n-1} w_{i_k i_{k+1}}}{\sum_{i_0} \cdots \sum_{i_n} P(i_0, ..., i_n) \prod_{k=0}^{n-1} w_{i_k i_{k+1}}} \tag{27}$$

Up to that point, it is most important to realize that no probabilistic aspect has been introduced so far. In other words, (27) can be viewed as a simple algebraic rewriting of (10). When we are able to compute exactly the sums involved in (27) both equations give the same answer. When this is possible we are in the realm of standard power methods and their improved versions (Lanczòs, Davidson,...). In that case we can refer to such approaches as "deterministic" power methods. If the sums are too large (for a linear space of size N there are $(N+1)^n$ terms in the sums and an iterative computation requires about nN^2 multiplications), some sort of "stochastic" power method (=QMC) is needed.

The essence of QMC is to reinterpret the quantity $P(i_0, ..., i_n)$ as the (n+1)-time probability distribution of a (Markov) chain defined by the transition probability distribution $P_{i \to j}$, (14). Once this is done, the various sums in equation (27) can be reinterpreted and computed as stochastic averages over the successive states drawn according to $P_{i \to j}$ (construction of a "stochastic trajectory"). In practice, the sequence of states is easily built on a computer as follows. For a given state i we first determine the set of states $\{j_1, \dots j_k \dots, j_K\}$ such that $\mathcal{P}_{j_k} \neq 0$ and the probability $\mathcal{P}_{j_k} = P_{i \to j_k}(\tau)$ for each state j_k is computed. A uniform random number ξ on the interval (0,1) is drawn and the state is chosen among this set according to the repartition function

$$\text{State } j_1 \text{ chosen if } \xi \in (0, \mathcal{P}_{j_1})$$

$$\text{State } j_2 \text{ chosen if } \xi \in (\mathcal{P}_{j_1}, \mathcal{P}_{j_1} + \mathcal{P}_{j_2})$$

$$...$$

$$\text{State } j_k \text{ chosen if } \xi \in \left(\sum_{l=0,k-1} \mathcal{P}_{j_l}, \sum_{l=0,k} \mathcal{P}_{j_l} \right)$$

$$...$$

(28)

This step can be iterated, thus giving rise to a chain of states, i.e, to a stochastic trajectory $i_0, i_1 \dots$. Finally, the estimate of the energy can be expressed as an average along the stochastic trajectory $(i_0, i_1, i_2, ...)$ as follows

$$E_0 = \frac{E_L(i_0) + E_L(i_1)w_{i_0 i_1} + E_L(i_2)w_{i_0 i_1}w_{i_1 i_2} + E_L(i_3)w_{i_0 i_1}w_{i_1 i_2}w_{i_2 i_3} + \dots}{1 + w_{i_0 i_1} + w_{i_0 i_1}w_{i_1 i_2} + w_{i_0 i_1}w_{i_1 i_2}w_{i_2 i_3} + \dots}$$

(29)

Note that passing from (27) to (29) is possible because the stochastic process used is recurrent or ergodic (see, e.g., Ref.[29]). As an important consequence, the quantity calculated in (29) does not depend neither on the particular stochastic trajectory used nor on the initial condition i_0. In other words, any stochastic trajectory of infinite length realizes the various transition probability distributions defined by the stochastic matrix.

Another point which deserves to be emphasized is that the stationary density of the stochastic process π_i is given by the squared components of the trial vector

$$\pi_i = v_i^2$$

(30)

The stationary density has a simple meaning: this is the distribution in configuration space which remains unchanged after application of the stochastic matrix.

In practice, this is the average distribution obtained on the computer when the equilibrium regime is reached. Mathematically, this condition is written as

$$\sum_i \pi_i P_{i \to j}(\tau) = \pi_j. \tag{31}$$

This relation is a simple consequence of the definition of the stochastic matrix (14).

In practice, it is found that a direct implementation of formula (29) is very inefficient. As the simulation time increases, no convergence of the energy is observed. The reason for that is the rapid increase of variance of the products of weights $\prod w_{i_l i_{l+1}}$ as a function of the number of iterations (number of weights). This important aspect is discussed in the next section.

2.4   Pure Diffusion Monte Carlo (PDMC)

In order to control the convergence of the energy as a function of the simulation time, the two limits $n \to \infty$ (number of iterations of the power method) and $P \to \infty$ (number of Monte Carlo steps for a given trajectory) must be treated *separately*. In other words, $E_0(n)$ is computed by collecting from the trajectory all contributions in (29) associated with n products of weights. Explicitly, it gives

$$E_0(0) = \frac{E_L(i_0) + E_L(i_1) + E_L(i_2) + \dots}{1 + 1 + 1 + \dots}$$

$$E_0(1) = \frac{E_L(i_1)w_{i_0 i_1} + E_L(i_2)w_{i_1 i_2} + E_L(i_3)w_{i_2 i_3} + \dots}{w_{i_0 i_1} + w_{i_1 i_2} + w_{i_2 i_3} \dots}$$

$$E_0(2) = \frac{E_L(i_2)w_{i_0 i_1}w_{i_1 i_2} + E_L(i_3)w_{i_1 i_2}w_{i_2 i_3} + E_L(i_4)w_{i_2 i_3}w_{i_3 i_4} + \dots}{w_{i_0 i_1}w_{i_1 i_2} + w_{i_1 i_2}w_{i_2 i_3} + w_{i_2 i_3}w_{i_3 i_4} \dots}$$

$$\dots \tag{32}$$

In practice it is found that, for not too large values of n (limited number of weights in the product $\prod w_{i_l i_{l+1}}$), $E_0(n)$ can be obtained with a good accuracy.

The method based on (32) is known in the literature as the Pure Diffusion Monte Carlo (PDMC) approach [29],[24]. Although intrinsically unstable at large n, PDMC has proven to be very useful as illustrated by a number of applications (*e.g.* [24],[35],[36], [37],[38],[39]). This is the case when the trial vector is accurate enough to allow the convergence of the various estimators before the large fluctuations at large iteration numbers arise.

Fig.3.1 presents for the toy model the convergence of $E_0(n)$ as a function of n. The parameters of the matrix H are identical to those used in the previous application and the trial vector chosen is $\mathbf{v} = (1/\sqrt{2}, 1/\sqrt{2})$. In order to get an estimate of the error made for a finite P at a given n, a number of independent calculations of $E_0(n)$ are made and the statistical errors are calculated using standard statistical techniques.

At $n = 0$ the Rayleigh quotient is given by

$$E_0(0) = \frac{\mathbf{v}^T H \mathbf{v}}{\mathbf{v}^T \mathbf{v}}. \tag{33}$$

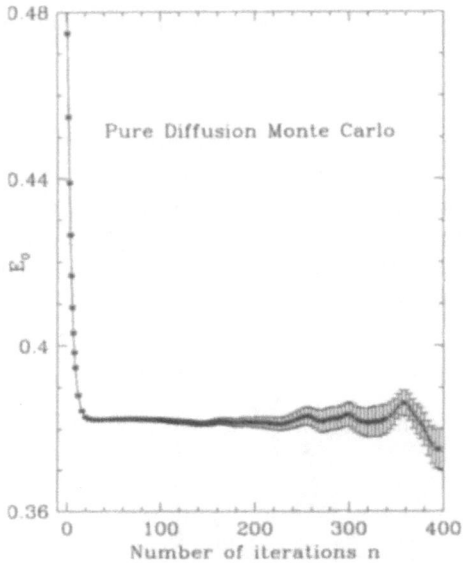

Figure 3.1: Pure Diffusion Monte Carlo calculation.

This quantity is also known as the "variational" energy associated with the trial vector. Here, the exact value for this quantity is 0.5 and we get 0.5001(2). When the iteration number n is increased the estimate of the energy converges to the exact value [number of iterations of about 40, $E_0(40) = 0.3822(2)$]. At larger values of n ($n > 100$) the difficulty associated with the problem of the many-weight product becomes severe and no stabilization is observed.

Note that in a very recent work a detailed mathematical analysis of the behavior of the PDMC algorithm as a function of the iteration number has been presented [40]. It has been shown in a rigorous way that the "effective" probability associated with a given state i as it is built in a PDMC scheme, namely

$$P_i = \sum_{P=0}^{\infty} \sum_{i_0} \cdots \sum_{i_{P-1}} \pi_{i_0} P_{i_0 \to i_1} \cdots P_{i_{P-1} \to i_P} w_{i_0 i_1} \cdots w_{i_{P-1} i} \qquad (34)$$

and considered as a random variable, does not converge as a function of P.

2.5   Diffusion Monte Carlo (DMC)

A most commonly employed method to escape from the difficulty associated with the weights at large n is to remove them from the averages and incorporate them in the stochastic process itself. This idea is realized by introducing a so-called birth-death or branching process. In practice, it consists in adding to the standard stochastic move of PDMC (28) a new step in which the current configuration is destroyed or copied a number of times proportional to the local weight. Denoting m the number of copies (multiplicity), we can take

$$m \equiv \text{int}(w_{ij} + \eta) \tag{35}$$

where $\text{int}(x)$ denotes the integer part of x, and η a uniform random number on $(0, 1)$. The use of (35) is justified because it gives the correct weight in average

$$\int_0^1 d\eta(w_{ij} + \eta) = w_{ij} \tag{36}$$

Note that, in contrast with PDMC which could be defined with one single walker, to simulate a branching process requires the use of a large enough population of walkers (in practice, a few thousands). Mathematically, adding a branching process can be viewed as sampling with a generalized transition probability $P_{i \to j}^*(\tau)$ defined as

$$P_{i \to j}^*(\tau) \equiv P_{i \to j}(\tau)w_{ij} = \frac{v_j}{v_i}[1 - \tau(H - E_T)]_{ij} \tag{37}$$

Of course, the normalization is no longer constant, $\sum_i P_{i \to j}^* \neq 1$ and is a function of j, and therefore P^* is not a genuine transition probability. However, we can still define a stationary density for it. From (37) it is easily shown that the stationary condition is obtained when E_T is chosen to be the exact energy E_0, and that the stationary density is now $v_i u_{0i}$

$$\sum_i \pi_i^* P_{i \to j}^* = \pi_j^* \tag{38}$$

with

$$\pi_i^* = v_i u_{0i} \quad \text{and} \quad E_T = E_0 \tag{39}$$

By using a stabilized population of configurations the exact energy may be now obtained as

$$E_0 = << E_L >> \equiv \lim_{P \to \infty} \frac{1}{P} \sum_{l=0}^{P-1} E_L(i_l) \tag{40}$$

where a population of states (walkers) is evolved according to (28) and branched according to (35). Such a Monte Carlo algorithm is known as the Diffusion Monte Carlo (DMC) method. Now, since in DMC the number of walkers can fluctuate, some sort of population control is required. Indeed, nothing prevents the total population from exploding or collapsing entirely. Various solutions to this problem have been proposed. Two common approaches consist either in performing from time to time a random deletion/duplication step or in varying slowly the reference energy E_T to keep the average number of walkers approximately constant. In both cases, a finite bias is introduced by the population control step. In order to minimize this undesirable source of error it is important to control the size of population as rarely as possible and in the most gentle way [1]. Here, to illustrate this idea, we do the population control by adjusting the reference energy to the fluctuations of population by using a formula of the type

$$E_T(t + \tau) = E_T(t) + K/\tau \ln[M(t)/M(t + \tau)] \tag{41}$$

where $M(t)$ is the total number of walkers at time t and K is some positive constant. When $M(t + \tau) > M(t)$ the reference energy is reduced and more walkers are killed at the next step. In the opposite case E_T is raised and more walkers are duplicated.

In Table 3.3 the DMC energy as a function of the size M of the population is shown. The bias for small population is clearly seen. For M sufficiently large the bias becomes smaller than the statistical fluctuations.

Table 3.3: Convergence of the energy in a DMC calculation as a function of the size M of the population for a fixed number of Monte Carlo steps. The total number of MC steps is about 2.10^7

Size M	$E_0(M)$
$M = 20$	0.3832(3.2)
$M = 40$	0.3829(3.2)
$M = 60$	0.3827(3.5)
$M = 80$	0.3822(3.2)
$M = 100$	0.3820(3.5)
$M = 200$	0.3820(3.4)
Exact value[a]	0.381966...

[a] See equation (2).

2.6   Stochastic reconfiguration Monte Carlo (SRMC)

Very recently, following an idea initially introduced by Hetherington [41], the use of stochastic reconfiguration processes has been reconsidered in Diffusion Monte Carlo [42],[43],[44],[40]. The motivation is to combine the best of both worlds: efficiency of DMC (most of the weights are treated in the stochastic process and the algorithm is stable) and absence of bias as in PDMC (no population control step). The approach is derived within a PDMC framework (the walkers "carry" some weight) but the population is "reconfigured" using some specific rules. The reconfiguration is done in a such way that the number of walkers is kept constant at each step and therefore no population control is required. The idea consists in carrying many walkers simultaneously and introducing a global weight associated with the entire population instead of a local weight for each walker. The global weight W of the population at iteration (or "time") l is defined as the average of the M local weights $w_{i_l i_{l+1}}^{(k)}$

$$W \equiv \frac{1}{M} \sum_{k=1}^{M} w_{i_l i_{l+1}}^{(k)} \tag{42}$$

where M is the number of walkers considered. By increasing the number M of walkers the fluctuations of the global weight is reduced and the gain in computational efficiency can be very important. Now, the method consists in defining a PDMC scheme in the enlarged linear space defined by the tensorial product (M times) of the initial linear space. In this new space the full stochastic matrix is defined as the tensorial product of individual stochastic matrices. Note that no correlation between the stochastic moves of different walkers is introduced at this level. Second, and this is the important point, each individual weight is rewritten as a function of the global weight

$$w_{i_l i_{l+1}}^{(k)} = \tilde{w}_{i_l i_{l+1}}^{(k)} W$$

with

$$\tilde{w}_{i_l i_{l+1}}^{(k)} \equiv \frac{w_{i_l i_{l+1}}^{(k)}}{W} \tag{43}$$

This re-writing allows to introduce the global weight as a weight common to all walkers and thus to define a standard PDMC scheme in the tensorial product of spaces. To take into account the new weight $\tilde{w}$ a so-called reconfiguration process is introduced. At each step the total population of M walkers is "reconfigured" by drawing M walkers independently with probability proportional to $\tilde{w}$. Note that, at this point, some correlation between different walkers is introduced since the duplication/deletion process for a walker depends on the global weight W.

In Fig.3.2 we present for our toy problem the calculation of the energy with the stochastic reconfiguration Monte Carlo method. A population of $M = 20$ walkers has been used. In sharp contrast with PDMC (see Fig.3.1) no problem at large values of n is observed. The converged value for the energy is 0.3819(6.4) is in excellent agreement with the exact value.

In order to improve further the algorithm it is possible to define a so-called optimal Stochastic Reconfiguration Monte Carlo method [40]. For that let us first discuss the two important limits of the algorithm, namely the case of an infinite number of walkers, $M \to \infty$ and the case of constant weights, $w \to 1$. When $M \to \infty$ the global weight (42) converges to its stationary exact value. As a consequence, the different weights $\tilde{w}$ associated with each walker (as given by (43)) become independent from each other and the reconfiguration process reduces to the usual branching process (35) without population control and systematic bias since the population is infinite. However, in the limit $w \to 1$ the method does not reduce to the standard DMC approach. Indeed, the reconfiguration step "reconfigures" the entire population whatever the values of the weights. In order to improve the efficiency of the SRMC method this undesirable source of fluctuations must be reduced. To do that we divide the population of walkers into two different sets. A first set of walkers corresponds to all walkers verifying $\tilde{w} \geq 1$. These walkers can be potentially duplicated and will be called "positive" walkers. The other walkers verify $0 \leq \tilde{w} < 1$, they can be potentially destroyed and will be called "negative walkers". The number of reconfigurations is defined as

$$N_{\text{Reconf}} = \sum_{k+} |\tilde{w}_{i_l i_{l+1}}^{(k)} - 1| = \sum_{k--} |\tilde{w}_{i_l i_{l+1}}^{(k)} - 1| \tag{44}$$

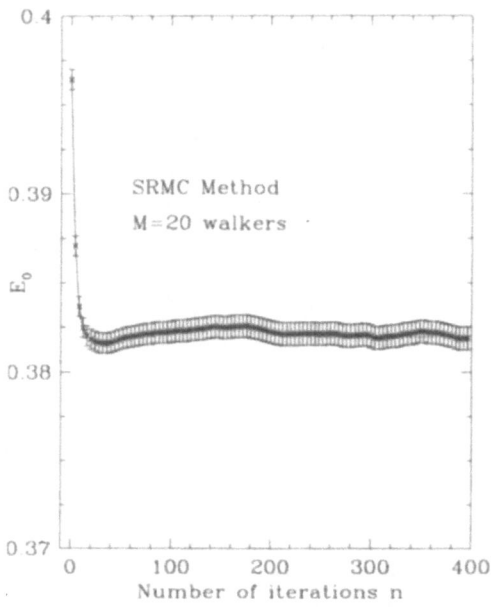

Figure 3.2: Stochastic Reconfiguration Monte Carlo calculation.

where $\sum_{k+}$ (resp. $\sum_{k-}$) indicates that the summation is done over the set of positive (resp. negative) walkers. The equality in (44) is a simple consequence of the definition of positive and negative walkers. Note that in general N_{Reconf} is not integer. An integer number of reconfigurations can be obtained by considering $\text{int}(N_{\text{Reconf}} + \eta)$ where η is a uniform random number on the interval (0,1). Once the number of reconfigurations has been drawn, a set of N_{Reconf} walkers is duplicated and another set of N_{Reconf} walkers is destroyed from the current population. This is done by drawing separately N_{Reconf} walkers among the lists of positive and negative walkers. It is easily verified that by doing this no source of systematic error has been introduced as in the original reconfiguration process of Hetherington. However, in contrast with the latter the average number of reconfigurations is kept minimal and, consequently, the efficiency of the simulation is significantly enhanced. In addition, the average number of reconfigurations vanishes as the weights become constant. In other words, the reconfiguration method reduces in this limit to the standard DMC method with $w = 1$.

In Table 3.4 we present the values obtained for the energy at zero time (without any correction due to the global weight), for the SRMC method and its optimal version. This quantity is particularly interesting since it is an indicator of the quality of the stationary density obtained. The closer the SRMC stationary density is to the DMC density (39), the smaller the difference between $E_0(n = 0)$ and $E_0(n = \infty) = E_0$ is. As it should be, both methods give an energy which converges

Table 3.4: Convergence of $E_0(n = 0)$ as a function of the size M of the population. The values for the standard Stochastic Reconfiguration Monte Carlo (SRMC) method and the optimal one are compared. The closer $E_0(n = 0)$ is to the exact result E_0 the closer the corresponding SRMC stationary density is to the exact DMC density, (39). The total number of MC steps is about 2.10^7

Size M	$E_0(n = 0)$ SRMC method	$E_0(n = 0)$ Optimal SRMC
$M = 10$	0.4091(5.7)	0.3940(4.7)
$M = 20$	0.3964(5.6)	0.3872(4.5)
$M = 100$	0.3848(5.7)	0.3830(4.1)
$M = 200$	0.3835(5.9)	0.3824(4.3)
$M = 400$	0.3829(5.7)	0.3822(4.4)
Exact valueE_0^a	0.381966...	0.381966...

[a] See equation (2).

to the exact value as the number of walkers goes to infinity. However, the finite bias at finite M is found to be much smaller for the optimal version. This result is a direct consequence of the fact that fluctuations due to the reconfiguration process are smaller in the optimal SRMC.

In Fig.3.3 the convergence of the energy as a function of the iteration number is shown. The curve is similar to what has been obtained previously with the standard SRMC method. However, for the same number of Monte Carlo steps the accuracy obtained is better. The converged value obtained for the energy is 0.3819(3.5).

2.7   Accelerating QMC

QMC calculations can be computationally demanding. Starting from a few thousands walkers, several thousands of Monte Carlo steps are performed for each walker. When the system studied is complex, this can indeed require an important numerical effort. In order to treat large systems and/or to get a better accuracy it is therefore desirable to accelerate as much as possible the convergence of Monte Carlo simulations. By "accelerating" it is meant to decrease significantly the statistical fluctuations without changing much the amount of calculations. To realize this, it is natural to imagine two main directions. A first direction consists in improving in some way the dynamics of the stochastic process. For example, configurations can be generated so that they are better distributed in configuration space or more independent. A second direction consists in decreasing as much as possible the fluctuations of the various estimators to be averaged. In what follows, we present two very recently proposed methods illustrating both aspects. Note that these approaches are of general interest (valid for almost any type of QMC methods).

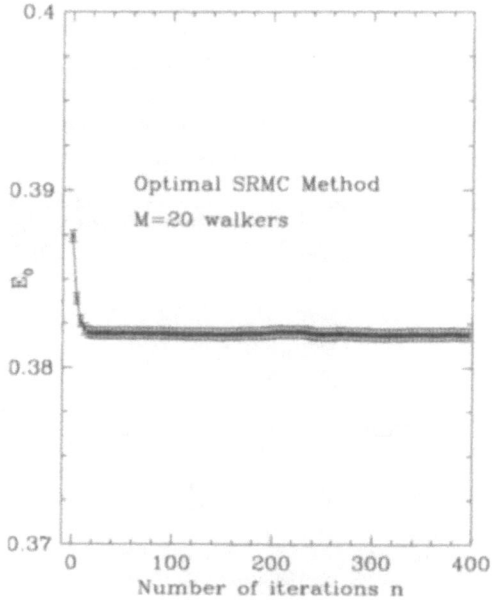

Figure 3.3: Optimal Stochastic Reconfiguration Monte Carlo calculation.

Use of integrated Poisson processes

In our toy example a stochastic trajectory can represented as a sequence of states 0 and 1 such as:

$$(0, 0, 0, 1, 0, 1, 1, 0, 0, 1, ...)$$

During a certain amount of "time" (or iterations) the system is blocked in a given state i (0 or 1) and then at some time escapes to a different state (here, $0 \to 1$ or $1 \to 0$). It is therefore possible to define the probability that the system remains in a given configuration i a number of times equal to n. It is given by

$$\mathcal{P}_i(n) \equiv P(i_1 = i, \tau; \ldots; i_{n-1} = i, \tau; i_n \neq i, \tau) =$$

$$[P_{i \to i}(\tau)]^n [1 - P_{i \to i}(\tau)] \tag{45}$$

$\mathcal{P}_i(n)$ defines a normalized discrete Poisson distribution

$$\mathcal{P}_i(n) = [1 - P_{i \to i}] \exp [n \ln(P_{i \to i})] \tag{46}$$

where the integer n runs from zero to infinity. It is quite easy to compute the average number of times the system is "trapped" in a given state i:

$$\bar{n}_i = \sum_{n=0}^{\infty} n \mathcal{P}_i(n) = \frac{P_{i \to i}(\tau)}{1 - P_{i \to i}(\tau)} \tag{47}$$

Now, to describe transitions towards states j different from i we introduce the following "escape" transition probability

$$\tilde{P}_{i \to j} = \frac{P_{i \to j}(\tau)}{1 - P_{i \to i}(\tau)} \quad j \neq i. \tag{48}$$

Using (14) $\tilde{P}_{i \to j}$ is rewritten in the most explicit form

$$\tilde{P}_{i \to j} = \frac{H_{ij} v_j}{\sum_{k \neq i} H_{ik} v_k} \quad j \neq i. \tag{49}$$

Once this "integrated" stochastic process is introduced the elementary stochastic trajectory is replaced by a trajectory between *different* states and to each state is associated an average "trapping time" given by

$$\bar{\theta}_i = \bar{n}_i \tau \tag{50}$$

All expressions presented previously can be re-expressed in terms of the "escape" transition probability and average trapping times only[45]. To give an example, within a DMC framework the energy estimate can be rewritten as:

$$E_0 = \frac{<< \bar{\theta}_i E_L(i) >>}{<< \bar{\theta}_i >>} \tag{51}$$

where the average $<< \dots >>$ is defined from a population of walkers evolving according to the escape transition probability and where the branching contribution associated with the transition from state i to state j is given now by the following "integrated" weight

$$\tilde{w}_i = \sum_{n=0}^{+\infty} \mathcal{P}_i(n) w_{ii}^n \tag{52}$$

This formula represents the average contribution of the weight w_{ii} when the system is blocked in state i.

Integrating out the entire time evolution of the system when blocked in a given state can improve a lot the convergence of QMC. This is particularly true when the system studied displays very different average "trapping" times (for example, when deep potential wells are present). Let us now illustrate this idea with some calculations. First, note that the application of the method to our 2×2 matrix is not representative of the general case. Indeed, for the 2×2 matrix, formula (51) degenerates to the ratio of two deterministic quantities (since the series of states is imposed: $...0 \to 1 \to 0 \to 1 \to 0...$). To test the method we need to consider a problem with at least three states. We have chosen a simple generalization of our toy model, namely

$$H = \begin{pmatrix} v_0 & -t & -t \\ -t & v_1 & -t \\ -t & -t & v_2 \end{pmatrix} \tag{53}$$

We take $t = 1, v_0 = 1$ and $v_1 = 2$. Here also, we consider a trial vector whose components are constant, $\mathbf{v} = (1/\sqrt{3}, 1/\sqrt{3}, 1/\sqrt{3})$. In Table 3.5 we present some calculations done with the standard DMC method and the improved version presented in this section. We consider the case of an intermediate value for v_2 ($v_2{=}3$) and the case of a larger value ($v_2{=}10$). Of course, the same number of Monte Carlo steps are performed in each case. When $v_2{=}3$ a clear improvement is observed. A factor of about 4 in statistical error has been gained. For the larger value, we get a factor of about 11. This important improvement is due to the fact that when v_2 is large the exact eigenvector is strongly localized on components 0 and 1 and therefore the on-site time integration is particularly effective.

Table 3.5: Integrating out the dynamics, $t = 1, v_0 = 1, v_1 = 2$.

Value of the potential v_2^a	$v_2 = 3$	$v_2 = 10$
E_0(standard)[b]	-0.21416(19)	0.18772(111)
E_0(integrated)[c]	-0.21430(4.4)	0.18805(10)
E_0(exact)	-0.2143197...	0.1880341...

[a] Defined by (53).
[b] Formula (40).
[c] Formula (51).

Improved estimators and zero-variance property

An alternative way of improving QMC methods consists in decreasing as much as possible the fluctuations of the quantity to be averaged. Since the ideas we shall present are in fact extremely general (valid for any type of Monte Carlo algorithm) let us work with very general notations. Let us call O an arbitrary observable defined in some discrete configuration space and π the normalized probability distribution of the Monte Carlo algorithm, we want to calculate the following average

$$< O >= \sum_i \pi_i O_i \qquad (54)$$

The magnitude of the statistical fluctuations associated with the use of a finite sample of configurations drawn according to π

$$< O >= \frac{1}{P} \sum_{i=1}^{P} O_i \qquad (55)$$

is directly related to the variance of the observable

$$\sigma(O)^2 \equiv < O^2 > - < O >^2. \qquad (56)$$

In order to accelerate the simulations we can use a so-called "improved" or "renormalized" estimator, that is, a new observable $\tilde{O}$ having the same average as the original one

$$< O >=< \tilde{O} > \qquad (57)$$

but a lower variance:

$$\sigma(\tilde{O}) < \sigma(O). \tag{58}$$

Various improved estimators have been introduced in the literature (See, e.g.,[46],[47]). The basic idea is to construct new estimators by integrating out some intermediate degrees of freedom and, therefore, removing the corresponding source of fluctuations. Very recently, we have proposed a general and systematic approach to construct such estimators[45]. The method is valid for any type of observable and any type of Monte Carlo algorithm. It is based on a zero-variance principle. The ideas are very simple and can be summarized as follows. We introduce a so-called trial matrix $\tilde{H}$ and a trial vector **u**. The basic requirements are that the trial matrix must be self-adjoint and must admit $\sqrt{\pi}$ as eigenstate with zero eigenvalue:

$$\sum_j \tilde{H}_{ij}\sqrt{\pi_j} = 0 \tag{59}$$

In the general case constructing a matrix verifying such conditions is not difficult. The "renormalized" or improved estimator is then defined as follows:

$$\tilde{O}_i = O_i + \frac{\sum_j \tilde{H}_{ij}u_j}{\sqrt{\pi_i}}. \tag{60}$$

It is quite easy to verify that both estimators have the same average. It is also easy to realize that the variance of $\tilde{O}$ will be minimal and equal to zero if the following fundamental equation is fulfilled:

$$\sum_j \tilde{H}_{ij}u_j = -(O_i - <O>)\sqrt{\pi_i} \tag{61}$$

This equation can be obtained by imposing the renormalized observable to be constant and equal to the exact average. Now, a very important point is that, whatever the choice of the trial matrix and trial vector, it is always possible to construct an improved estimator having a lower variance than the bare observable O. The proof is as follows. Let us introduce α a multiplicative factor for the trial vector **u** and let us minimize the variance of the renormalized estimator with respect to this parameter. We write:

$$\sigma(\alpha)^2 = <\tilde{O}^2> - <\tilde{O}>^2 = \sigma(\alpha=0)^2 + 2\alpha <OHu/\sqrt{\pi}> + \alpha^2 <(Hu/\sqrt{\pi})^2> \tag{62}$$

Minimizing the variance with respect to α we get

$$\alpha_{opt} = - <OHu/\sqrt{\pi}> / <(Hu/\sqrt{\pi})^2> \tag{63}$$

and

$$\sigma(\alpha_{opt})^2 = \sigma(\alpha=0)^2 - <OHu/\sqrt{\pi}>^2 / <(Hu/\sqrt{\pi})^2> \tag{64}$$

The correction to $\sigma(\alpha=0)^2$ being negative, we see that from *any* trial vector we can get a better estimator just by optimizing the multiplicative factor. Applications of the zero-variance property to a number of classical and quantum Monte Carlo simulations have shown that very important gains in computational effort can be obtained[48].

3   QMC beyond the 2×2 matrix

All ideas and results presented in the preceding sections can be extended to a general matrix or a continuous operator. In this section we discuss a number of aspects related to the general case. For the sake of clarity we shall discuss separately the cases of a discrete or a continuous configuration space.

3.1   QMC for a $N \times N$ matrix

For a finite matrix of arbitrary dimension the various formulas presented above can be readily used (most of the formulas have already been written for an arbitrary size N of the linear space). For our 2×2 matrix we have restricted ourselves to the case of a negative off-diagonal element, $H_{12} = -t < 0$. It was justified since it corresponds to choose arbitrarily the overall sign of the matrix. In the general case ($N > 2$), off-diagonal elements H_{ij} can now have a positive or negative sign. This fact has to be taken into account when defining the stochastic matrix. Let us recall that the value of the stochastic matrix between two different states is given by (see, (14))

$$P_{i \to j}(\tau) = -\frac{v_j}{v_i} \tau H_{ij} \quad i \neq j \tag{65}$$

To allow a probabilistic interpretation this quantity needs to be positive. As explained in the introduction this is true for a number of quantum systems describing bosons or distinguishable particles. For fermions, both H_{ij} and the trial vector components have a non-constant sign. As already discussed a number of practical solutions have been proposed to solve this famous "sign problem". We shall not discuss further this point and we refer the interested reader to the abundant literature on this subject.

Now, it is worth emphasizing that the quantity which controls the efficiency of QMC is the trial vector used, $\mathbf{v}$. When $\mathbf{v}$ is close to the exact eigenvector, the sampling of the configuration space is done efficiently and the statistical fluctuations on the energy can be very small. When the trial vector is good enough, the size of the linear space treatable can be almost as large as desired (virtually infinite). Many QMC applications on lattice models have illustrated this property (see, e.g. [3]). To give a concrete example, in a recent study of a metal-insulator phase transition in a generalized one-dimensional Hubbard model[45], the size of the largest matrix to be diagonalized was $N \sim 1.2 \ 10^{28}$ (32 particles moving on a 32 sites-ring). For this matrix and using some physically meaningfull trial vectors the values obtained for the ground-state energy were: $E_0 = -52.13056(15)$ in the better case (small electron-electron repulsion) and $E_0 = -23.7118(13)$ in the worst case (large electron-electron repulsion).

3.2   QMC in a continuum

For continuous configuration spaces all ideas are identical but their implementation can appear in a quite different way. Since the Schrödinger Hamiltonian has a simple form in the x-representation, QMC is developped within this representation. The fundamental step of the power method is written as

$$\psi^{(n+1)}(\mathbf{x}) = \int d\mathbf{x}'[G(H)]_{\mathbf{xx}'}\psi^{(n)}(\mathbf{x}') \tag{66}$$

where $[G(H)]_{\mathbf{xx}'}$ denotes the kernel of the operator $G(H)$ (matrix elements in a continuous space), $\mathbf{x}$ represents a vector in the d-dimensional space $\mathbf{x} = (x_1, ..., x_d)$ and $\psi^{(n)}(\mathbf{x})$ is the n-th order iterated trial function. The vast majority of QMC applications have been concerned with a Hamiltonian operator of the form

$$H = -\frac{1}{2}\sum_{i=1}^{d}\frac{\partial^2}{\partial x_i^2} + V(x), \tag{67}$$

that is, the standard quantum-mechanical Hamiltomian used to describe a set of interacting particles. The first part in (67) represents the kinetic energy and $V(x)$ is the potential function. For such an operator the kernel of H is highly singular (derivatives of Dirac distributions). Therefore, in contrast with what is done for finite linear spaces, the operator $G(H) = 1 - \tau(H - E_T)$ cannot be used and some regular kernels must be introduced. The most common choice is

$$G(\mathcal{H}) = e^{-\tau(\mathcal{H}-E_T)}. \tag{68}$$

For continuous systems the QMC approach based on (68) is in general referred to as Diffusion Monte Carlo (note that several denominations exist in the literature, we give here what we think to be the most current one). Another possible choice is

$$G(\mathcal{H}) = \frac{1}{1 + \tau(\mathcal{H} - E_T)}. \tag{69}$$

The main advantage of this second choice is to define a QMC method free of the so-called short-time approximation (see below). The methods based on (69) are usually referred to as Green's function Monte Carlo[1],[10].

In the same way as in the discrete case we can define a stochastic kernel (matrix) or transition probability distribution as follows:

$$P(\mathbf{x} \to \mathbf{y}, \tau) = \frac{\psi_T(\mathbf{y})}{\psi_T(\mathbf{x})}[e^{-\tau(H-E_L)}]_{\mathbf{xy}} \tag{70}$$

where $\psi_T(\mathbf{x})$ is some trial function and E_L is the local energy function

$$E_L(\mathbf{x}) = H\psi_T(\mathbf{x})/\psi_T(\mathbf{x}). \tag{71}$$

This definition is the direct generalization of (14) and (16). Note that $P(\mathbf{x} \to \mathbf{y}, \tau) \geq 0$ and that $\int d\mathbf{y}P(\mathbf{x} \to \mathbf{y}, \tau) = 1$. In order to sample $P(\mathbf{x} \to \mathbf{y}, \tau)$ we need to interpret it as a probability distribution with respect to $\mathbf{y}$ for a fixed value of $\mathbf{x}$. This can be done in a natural way since $P(\mathbf{x} \to \mathbf{y}, \tau)$ obeys the following Fokker-Planck equation

$$\frac{\partial P}{\partial \tau} = LP \tag{72}$$

where L is a Fokker-Planck operator

$$L = \frac{1}{2}\nabla^2. - \nabla[\mathbf{b}.]\qquad(73)$$

and $\mathbf{b}$ the so-called drift vector given by

$$\mathbf{b} = \frac{\nabla\psi_{\mathbf{T}}}{\psi_T}.\qquad(74)$$

The derivation of this result can be found in Ref.[29]. For a presentation of the Fokker-Planck equation and material related to diffusion processes, the reader is referred to [49]. In a continuum $P(\mathbf{x}\to\mathbf{y},\tau)$ is therefore interpreted as the transition probability density of a standard drifted continuous diffusion process in configuration space. Stochastic trajectories can be obtained by using the Langevin equation associated with the diffusion process[49]

$$d\mathbf{x}(t) = \mathbf{b}[\mathbf{x}(t)]dt + d\mathbf{W}(t)\qquad(75)$$

where $\mathbf{W}(t)$ is the standard multi-dimensional Wiener process. In practice, this equation is discretized under the form

$$\mathbf{x}(t+\tau) = \mathbf{x}(t) + \mathbf{b}[\mathbf{x}(t)]\tau + \sqrt{\tau}\eta\qquad(76)$$

where η is a vector whose components are independent gaussian random variables verifying $<\eta_i> = 0$ and $<\eta_i\eta_j> = \delta_{ij}$. This latter equation is particularly simple to implement on a computer. This is the continuous counterpart of (28) used in the discrete case. Note that the discretization in time introduces a bias (short-time approximation). To obtain exact quantities several simulations with different time-steps are performed and the results are then extrapolated to a zero time-step. Note that this source of error does not exist in the discrete case.

Finally, the weights can be defined via the fundamental relation between the operator $G(H)$ and the stochastic kernel. We can write

$$\frac{\psi_T(\mathbf{y})}{\psi_T(\mathbf{x})}[e^{-\tau(H-E_T)}]_{\mathbf{xy}} = \frac{\psi_T(\mathbf{y})}{\psi_T(\mathbf{x})}[e^{-\tau[(H-E_L)+(E_L-E_T)]}]_{\mathbf{xy}}.\qquad(77)$$

In the short-time approximation (or "Trotter break up", see [50]) this equation can be rewritten as

$$\frac{\psi_T(\mathbf{y})}{\psi_T(\mathbf{x})}[e^{-\tau(H-E_T)}]_{\mathbf{xy}} \sim_{\tau\to0} P(\mathbf{x}\to\mathbf{y},\tau)e^{-\tau[E_L(\mathbf{x})-E_T]}\qquad(78)$$

and the weight appears now as a simple multiplicative operator given by

$$w(\mathbf{x}) = e^{-\tau[E_L(\mathbf{x})-E_T]}.\qquad(79)$$

Using expression (70) for the transition probability distribution [stochastically realized via (76)] and expression (79) for the weight, all ideas presented in Sec. II can be implemented.

Acknowledgements. We would like to acknowledge many discussions with Anatole Khelif (Paris 7) and some useful comments on the manuscript from Peter Reinhardt (Paris 6). This work was supported by the "Centre National de la Recherche Scientifique" (CNRS). Part of the work has been done by R. Assaraf at SISSA under EU sponsorship, TMR Network FULPROP.

Bibliography

[1] D.M. Ceperley and M.H. Kalos, in *Monte Carlo Method in Statistical Physics*, edited by K.Binder (Springer-Verlag, Heidelberg, 1992).

[2] D.M. Ceperley, Rev. Mod. Phys. **67**, 279 (1995).

[3] W. von der Linden, Phys. Rep. **220**, 53 (1992).

[4] B.L. Hammond, W.A. Lester,Jr., and P.J. Reynolds in *Monte Carlo Methods in Ab Initio Quantum Chemistry*, World Scientific Lecture and course notes in chemistry Vol.1 (1994).

[5] J.B. Anderson in *Reviews in Computational Chemistry*, Vol.13, edited by K.B. Lipkowitz and D.B. Boyd (Wiley, 1999).

[6] A.R. Gourlay and G.A. Watson, *Computational Methods for Matrix Eigenproblems*, Chichester: Wiley.

[7] C. J. Umrigar, K. G. Wilson, and J. W. Wilkins, Phys. Rev. Lett. **60**, 1719 (1988).

[8] D.M. Ceperley, J. Stat. Phys. **63**, 1237 (1991).

[9] K.E. Schmidt amd M.H. Kalos in *Monte Carlo methods in Statistical Physics II*, edited by K. Binder, Springer, (1984).

[10] D.M. Ceperley and B.J. Alder, J. Chem. Phys. **64**, 5833 (1984).

[11] J.B.Anderson in *Understanding Chemical Reactivity*, S.R. Langhoff, Ed. Kluwer (1995).

[12] J.B. Anderson, Int. Rev. Phys. Chem. **14**, 95 (1995).

[13] K.E Schmidt and J.W. Moskowitz, J. Chem. Phys. **93**, 4172 (1990).

[14] M.P. Nightingale and C.J. Umrigar in *Recent Advances in Computational Chemistry* Vol.2, edited by W.A. Lester Jr., World Scientific (1997).

[15] T. Yoshida and K. Igushi, J. Chem. Phys. **88**, 1032 (1988).

[16] G.B. Bachelet, D.M. Ceperley, and M.G.B. Chiocchetti, Phys. Rev. Lett **62**, 2088 (1989).

[17] G.B. Bachelet, D.M. Ceperley, M. Chiocchetti, and L. Mitas, "Atomic pseudo-hamiltonians for quantum Monte Carlo", Dordrecht, Kluwer (1989)

[18] B.L. Hammond, P.J. Reynolds, and W.A Lester, Jr., Phys. Rev. Lett **61**, 2312 (1988).

[19] B.L. Hammond, P.J. Reynolds, and W.A Lester, Jr., J. Chem. Phys. **87**, 1130 (1987).

[20] L. Mitas, E.L. Shirley, and D.M. Ceperley, J. Chem. Phys. **95**, 3467 (1991).

[21] M.M. Hurley and P.A. Christiansen, J. Chem. Phys. **86**, 1069 (1987).

[22] E. Clementi and C. Roetti, *Atomic Data and Nuclear Data Tables* **14**, 177(1974)

[23] K. Frankowski and C.L. Pekeris, Phys. Rev. **146**, 46 (1966).

[24] M. Caffarel and P. Claverie, J. Chem. Phys. **88** 1100 (1988).

[25] W. Kolos and L. Wolniewicz, J. Chem. Phys. **43**, 2429 (1965).

[26] B.N. Parlett, *The symmetric Eigenvalue Problem, Prentice-Hall Series in Computational Mathematics* (1980).

[27] M. Caffarel, F.X. Gadea, and D.M. Ceperley, Europhys. Lett. **16**, 249 (1991).

[28] E.R. Davidson, J. Comput. Phys. **17**, 87 (1975)

[29] M. Caffarel and P. Claverie, J. Chem. Phys. **88** 1088 (1988).

[30] M. Caffarel and D.M. Ceperley, J. Chem. Phys. **97** 8415 (1992)

[31] B. Chen and J.B. Anderson, J. Chem. Phys. **102**, 4491 (1995).

[32] C. Filippi and C.J. Umrigar, J. Chem. Phys. **105**, 213 (1996).

[33] J.C. Grossman, L. Mitas, and K. Raghavachari, Phys. Rev. Lett. **75**, 3870 (1995).

[34] J.C. Grossman, and L. Mitas, Phys. Rev. Lett. **74**, 1323 (1995).

[35] K.J. Runge, Phys. Rev. B **45**, 7229 (1992).

[36] D.M. Ceperley and B. Bernu, J. Chem. Phys. **89** 6316 (1988).

[37] B. Bernu, D.M. Ceperley, and W.A. Lester Jr., J. Chem. Phys. **93** 552 (1990).

[38] M. Caffarel, M. Rérat, and C. Pouchan, Phys. Rev. A **47**, 3704 (1993).

[39] F. Schautz and H.J. Flad, J. Chem. Phys. **110** 11700 (1999).

[40] R. Assaraf, M. Caffarel, and A. Khelif, "Diffusion Monte Carlo with a fixed number of walkers", Phys. Rev. E (April 2000)

[41] J.H. Hetherington Phys. Rev. A **30**, 2713 (1984).

[42] S. Sorella, Phys. Rev. Lett. **80**, 4558 (1998).

[43] M. Calandra Buonaura and S. Sorella, Phys. Rev. B **57**, 11446 (1998).

[44] S. Sorella and L. Capriotti, Phys. Rev. B **61**, 2599 (2000).

[45] R. Assaraf, P. Azaria, M. Caffarel, and P. Lecheminant, Phys. Rev. B **60**, 2299 (1999).

[46] M. Sweeny, Phys. Rev. B **27**, 4445 (1983); G. Parisi, R. Petronzio, and F. Rapuano, Phys. Lett. B **128**, 418 (1983); U. Wolff, Nucl. Phys. B **334**, 581 (1990)

[47] B. Ammon, H.G. Evertz, N. Kawashima, M. Troyer, and B. Frischmuth, Phys. Rev. B **58**, 4304 (1998).

[48] R. Assaraf and M. Caffarel, Phys. Rev. Lett. **83** , 4682 (1999)

[49] C.W. Gardiner, *Handbook of Stochastic Methods* (Springer, Berlin, 1983).

[50] H.F. Trotter, Proc. Am. Math. Soc. **10**, 545 (1959); M. Suzuki, Commun. Math. Phys. **51**, 183 (1976).

Chapter 4

On the controllability of bilinear quantum systems

Gabriel Turinici
ASCI-CNRS Laboratory,
Bât. 506,
Université Paris Sud,
F-91405 Orsay Cedex
turinici@asci.fr

Abstract: We present in this paper controllability results for quantum systems interacting with lasers. A negative result for infinite dimensional spaces serves as a starting point for a finite dimensional analysis. We show that under physically reasonable hypothesis in such systems we can control the population of the eigenstates. Applications are given for a five-level system.

1   Introduction

Controlling chemical reactions at the quantum level was a long-lasting goal for the Chemists (cf. [2], [6], [11], [12], [14], [15], [16], [19]) from the very beginning of the laser technology. Indeed, due to the subtle nature of the interactions involved, this kind of manipulation is expected to allow on the one hand for much efficient and finer control than classical tools (temperature, pressure, catalyzers ...) and on the other hand for new reactions and/or products to be obtained.

The first experiments have shown that designing the laser pulse able to steer the system to the desired target state is a rather difficult task that physical intuition alone cannot accomplish. It is only recently that tools coming from the control theory began to give satisfactory results in some particular cases; finding the optimal electric field is now treated by numerical methods and new models are sought after that be also reliable and cheap from a computational point of view.

A legitimate question arises in this context: what quantum states can be attained using such an external field ? Some answers are given below.

2   Infinite dimensional controllability

Our purpose is to control the equations that govern the time evolution of quantum systems. Let consider such a system (isolated from the outer world for the moment) whose internal Hamiltonian is H_0 that is prepared in the initial state $\Psi_0(x)$; its dynamics obeys the Time Dependent Schrödinger Equation. Denoting by $\Psi(x,t)$ the state at the time t one can write the evolution equations for the free system:

$$\begin{cases} i\hbar\frac{\partial}{\partial t}\Psi(x,t) = H_0\Psi(x,t) \\ \Psi(x,t=0) = \Psi_0(x), \ \|\Psi_0\|_{L^2(\mathbf{R}^\gamma)} = 1 \end{cases} \tag{1}$$

In the presence of external interactions that for us will be an electric field created by a laser and modeled by a laser intensity $\epsilon(t) \in \mathbf{R}$ and by a certain time independent dipole moment operator[1] $\mathcal{B}$ the (controlled) dynamical equations reads:

$$\begin{cases} i\hbar\frac{\partial}{\partial t}\Psi_\epsilon(x,t) = H_0\Psi_\epsilon(x,t) - \epsilon(t)\mathcal{B}\Psi_\epsilon(x,t) = H\Psi_\epsilon(x,t) \\ \Psi_\epsilon(x,t=0) = \Psi_0(x) \end{cases} \tag{2}$$

The goal is to find (if any) a final time T and a finite energy laser pulse $\epsilon(t)$, $\epsilon(t) \in L^2([0,T])$ able to steer the system from $\Psi_0(x)$ to some predefined target $\Psi_\epsilon(x,T) = \Psi_{target}(x)$.

Note that $\Psi_\epsilon(x,t)$ is evolving on the unit sphere $S(0,1)$ of $L^2(\mathbf{R}^\gamma)$:

$$S(0,1) = \{f \in L^2(\mathbf{R}^\gamma); \|f\|_{L^2(\mathbf{R}^\gamma)} = 1\}$$

---

[1] Of course, depending on the problem at hand, one may sometime choose to go beyond this first-order, bilinear term when describing the interaction between the laser and the system, cf. [7], [8].

Indeed one can easily prove that the L^2 norm of Ψ_ϵ is conserved throughout the evolution:

$$\|\Psi_\epsilon(x,t)\|_{L^2_x(\mathbf{R}^\gamma)} = \|\Psi_0\|_{L^2(\mathbf{R}^\gamma)}, \ \forall t > 0. \tag{3}$$

Let us point out some simple (but important) remarks before carrying on the analysis of these equations. Firstly in what the target state is concerned it follows by the incertitude principle that one will never be able to experimentally verify, neither exploit, the exact controllability. In fact even if one method gives exactly the desired target state Ψ_{target} the free evolution (i.e. when laser is switched off $\epsilon(t) = 0, t \geq T$) of the quantum system **instantaneously modifies** this state (by a time dependent phase factor if Ψ_{target} is an eigenfunction of H_0 and by the (1) formula in general).

In this context a first negative controllability result is therefore not really restrictive. In fact using compacity arguments as those in [1] we can prove the following[2]:

Theorem 1 *Let $\mathcal{B}$ be a bounded operator from $H^2_x(\mathbf{R}^\gamma)$ to itself and let H_0 generate a C^0 semigroup of bounded linear operators on $H^2_x(\mathbf{R}^\gamma)$. Denote by $\Psi_\epsilon(x,t)$ the solution of (2). Then the set of attainable states from Ψ_0 defined by*

$$\mathcal{AS} = \cup_{T>0}\{\Psi_\epsilon(x,T); \epsilon(t) \in L^2([0,T])\} \tag{4}$$

is contained in a countable union of compact subsets of $H^2_x(\mathbf{R}^\gamma)$. In particular its complement with respect to $S(0,1)$: $\mathcal{N} = S(0,1) \setminus \mathcal{AS}$ is everywhere dense on $S(0,1)$. The same holds true for the complement with respect to $S(0,1) \cap H^2_x(\mathbf{R}^\gamma)$.

Proof. To prove the first part of the theorem one applies Thm. 3.6 from [1] on the space $H^2_x(\mathbf{R}^\gamma)$ for the operators $-iH_0$ and $-i\mathcal{B}$ (and restricts $\epsilon(t)$ to L^2 functions).

Note that for any compact subset K of X

$$[0,n] \cdot K = \{rf; 0 \leq r \leq n, f \in K\}$$

is also compact. Applying this to the compact components K of $\mathcal{AS}$ one notes that

$$\cup_{r \geq 0} r\mathcal{AS} = \cup_{n \in \mathbf{N}} [0,n] \cdot \mathcal{AS}$$

is also a countable union of compacts subsets of $H^2_x(\mathbf{R}^\gamma)$. It follows by the Baire category theorem that $\cup_{r \geq 0} r\mathcal{AS}$ has dense complement in $H^2_x(\mathbf{R}^\gamma)$; in particular the complement of $\mathcal{AS}$ with respect to $S(0,1) \cap H^2_x(\mathbf{R}^\gamma)$ has to be everywhere dense on $S(0,1) \cap H^2_x(\mathbf{R}^\gamma)$.

Given this result one may either study the controllability with respect to a finite number of moments or the controllability of the corresponding finite dimensional system. Is the second analysis that we chose to pursue in this paper.

---

[2]We refer to [10] for a different view on this issue. Let us point out however that their analysis is done on piecewise constant functions which may not always carry physical meaning for our problem; in particular one may prove controllability in this class but realize (by the theorem we present here) that this controllability requires infinite L^2 norm and therefore infinite laser energy.

3 Finite dimensional controllability

Let then $D = \{\Psi_i(x); i = 1, .., N\}$ be an orthonormal basis for a finite dimensional sub-space of $L^2(\mathbf{R}^\gamma)$ that we are interested in[3] and A and B be the matrices of the operators H_0 and $\mathcal{B}$ respectively, with respect to this base[4].

Let us denote by $C = (c_i)_{i=1}^N$ the coefficients of $\Psi_i(x)$ in the formula of the evolving state $\Psi(t, x) = \sum_{i=1}^N c_i(t)\Psi_i(x)$; then the equations (2) read

$$\begin{cases} i\hbar \frac{\partial}{\partial t}C_\epsilon = AC_\epsilon - \epsilon(t)BC_\epsilon \\ C_\epsilon(t = 0) = C_0 \end{cases} \tag{5}$$

$$C_0 = (c_{0i})_{i=1}^N, \ c_{0i} = \int_{\mathbf{R}^\gamma} \Psi_0 \Psi_i dx \tag{6}$$

The controllability of (5) has been dealt with in the literature (cf. [13]) by considering the problem of the controllability of a system posed on the space of the unitary matrices of dimension N. This approach has the benefit of granting us access to the general tools and results on bilinear controllability on Lie groups. However it does not correspond to a physical necessity and therefore gives criterions not so easy to verify; moreover all the results one can obtain this way give only sufficient conditions for exact controllability (due to the setting that is more general). Finally there exists a class of simple quantum systems controllable (in a sense to be defined further on) that do not verify the criteria emerging from the Lie group analysis.

We have therefore judged instructive to study this issue taking into account the specificity of the quantum framework; we were thus lead into identifying **necessary and sufficient** conditions for the finite dimensional controllability[5].

In the case of our modeling[3] [4] the A matrix is diagonal and B is symmetrical and has null diagonal elements[6]. Let us denote by λ_i, $i = 1, .., N$ the diagonal elements of A (the energies of the states Ψ_i).

Before presenting our controllability results we have to introduce the first elements required to explain our controllability concept. As it was previously seen the system evolves on the unit sphere of $L_x^2(\mathbf{R}^\gamma)$ which in finite dimensional representation reads:

$$\sum_{i=1}^N |c_{\epsilon i}(t)|^2 = 1, \ \forall t \geq 0 \tag{7}$$

From (5) one notices that when the system is evolving freely ((5) with $\epsilon(t) = 0$) the (relative) phases of the coefficients $c_{\epsilon i}(t)$ change but not the populations of the eigenstates. We will therefore study only the population transfer between

---

[3] This space is given by our model and the functions $\Psi_i(x)$ are usually the first eigenfunctions of H_0 constructed by a prior computation or by a modeling based on observations.

[4] We suppose in the begining that $\mathcal{B}$ is such that $B_{ii} = 0$, $i = 1, .., N$; for the general case see the appendix.

[5] see also [4] for an overview of the topic

[6] see the appendix for the general case

eigenstates[7] i.e. only changes in $|c_i(t)|^2$. We call *population distribution* for the system (5) any N-tuple $d \in \mathbf{R}^N$ such that

$$\sum_{i=1}^{N} d_i^2 = 1, \ d_i \geq 0, \ i = 1, ..., N \tag{8}$$

We will also say that we can *reach the population distribution d from the initial state C_0* if for any $\eta > 0$ there exists a final time $T_d > 0$ and an electric field $\epsilon(t) \in L^2([0, T_d])$ such that the solution of (5) satisfy $||c_{\epsilon k}(T_d)|^2 - d_k^2| < \eta$. If this is also true for $\eta = 0$ then we say that we can *exactly reach the population distribution d from the initial state C_0*.

4 Transfer graph and necessary conditions

According to the physical intuition that we will support in the following by mathematical arguments, the B matrix describes the population flow among different eigenstates of the system. In order to formalize this idea we associate to the system some graph $G = (V, E)$ called the *transfer graph*. We define the set V of vertices as the set of eigenstates Ψ_i and the set of edges E as the set of all pairs of eigenstates coupled by the matrix B. Since B is symmetrical we can consider G non-oriented:

$$G = (V, E): \quad V = \{\Psi_1, ..., \Psi_n\} \ \ E = \{(\Psi_i, \Psi_j); B_{ij} \neq 0\} \tag{9}$$

Let us decompose this graph into connected components $G_\alpha = (V_\alpha, E_\alpha)$, $\alpha = 1, .., K$. Note that this decomposition corresponds to a bloc-diagonal structure of the matrix B (modulo some permutations on the indices). Using this decomposition one can write new conservation laws for each connected component:

$$\sum_{\{i; \Psi_i \in V_\alpha\}} |c_{\epsilon i}(t)|^2 = constant, \ \forall t > 0, \ \alpha = 1, .., K \tag{10}$$

In order to justify (10) one checks by the definition of G and using equations (5) that for all $\alpha = 1, .., K$:

$$i\hbar \frac{\partial}{\partial t} \sum_{\{i; \Psi_i \in V_\alpha\}} |c_{\epsilon i}(t)|^2 = 0 \tag{11}$$

This allows us to give necessary conditions for controllability:

Lemma 1 *If one can reach the population distribution d from the initial configuration C_0 then*

$$\sum_{\{i; \Psi_i \in V_\alpha\}} |c_{0i}|^2 = \sum_{\{i; \Psi_i \in V_\alpha\}} d_i^2, \ \alpha = 1, .., K \tag{12}$$

Definition 1 *We say that* **the population distribution of the system (5) is controllable** *if for any initial state C_0 and any population distribution d that satisfy (12) it is possible to reach d from the initial configuration C_0.*

---

[7]we refer the reader to [18] for more general controllability results concerning the state of the system

5   Controllability results

Denote $\omega_{kl} = \lambda_k - \lambda_l$, $k, l = 1, ..., N$. To ease the notations we will be working in atomic units ($\hbar = 1$). Let us introduce the following hypothesis:

H The components G_α, $\alpha = 1, .., K$ of G remain connected after elimination of all edge pairs $(\Psi_i, \Psi_j), (\Psi_a, \Psi_b)$ such that $\omega_{ij} = \omega_{ab}$ (*degenerate transitions*).

5.1   Local exact controllability

Theorem 2 *Let d_0 be the population distribution associated to the initial state C_0: $d_0 = (|c_{0i}|)_{i=1,...,N}$. Suppose $d_{0i} \neq 0$, $i = 1, ..., N$ and that the hypothesis (H) is verified. Then there exists an open neighborhood D of d_0 on the surface of $\mathbf{R}^N$ given by the necessary conditions (12) endowed with the canonical topology such that one can exactly reach any population distribution d in D from C_0.*

Remark 1 *The conditions $d_{0i} \neq 0$, $i = 1, ..., N$ are just technicalities needed in the proof. Note that if some $d_{0i} = 0$ one has to take care when choosing the good target set to expect exact controllability into, since there is no reason to hope in (exactly) reaching population "distributions" having* **strictly negative** *population in some eigenstates. This is indeed the part that makes things more involved.*

Proof. In order to better highlight the key elements of the proof we treat only the case $\omega_{ij} \neq \omega_{ab}$, $\forall (i, j) \neq (a, b)$, the general case bearing no new concepts. Let us denote $\overline{A} = -iA$ and $\overline{B} = -iB$. Then (5) become:

$$\begin{cases} \frac{\partial}{\partial t} C_\epsilon = (\overline{A} + \epsilon(t)\overline{B})C_\epsilon \\ C_\epsilon(t = 0) = C_0 \end{cases} \tag{13}$$

Denote by $c(\epsilon, C_0, t) = (c_a(\epsilon, C_0, t))_{a=1}^N$ the solution at the time t of (13) for the initial ($t = 0$) data C_0 and electric field $\epsilon(t)$.

We define the application $M : L^2(\mathbf{R}) \times \mathbf{R} \to \mathbf{R}^N$ given by

$$M(\epsilon, t) = (|c_a(\epsilon, C_0, t)|^2)_{a=1}^N \tag{14}$$

Note that by the necessary conditions (12) the range of M is a subset of

$$\{(x_i)_{i=1}^N \in \mathbf{R}^N; \sum_{\{i; \Psi_i \in V_\alpha\}} x_i = \sum_{\{i; \Psi_i \in V_\alpha\}} |c_{0i}|^2, \ \alpha = 1, .., K\}$$

The system (13) can be written in the integral form:

$$c(t) = e^{\overline{A}t}c(0) + \int_0^t e^{\overline{A}(t-s)}\epsilon(s)\overline{B}c(s)ds \tag{15}$$

which gives us (cf. also [1]) the formula of the (Fréchet) derivative $D_\epsilon c(\epsilon, C_0, t)$ of $c(\epsilon, C_0, t)$ with respect to ϵ computed at $\epsilon(t) \equiv 0$:

$$D_\epsilon c(\epsilon, C_0, t)|_{\epsilon=0} \cdot \tilde{\epsilon} = \int_0^t e^{\overline{A}(t-s)}\tilde{\epsilon}(s)\overline{B}e^{\overline{A}s}c(0)ds \tag{16}$$

Denoting by $w(t)$ the free evolution of the system ($w_a(t) = c_a(0, C_0, t)$ and $w(t) = c(0, C_0, t)$) and using the canonical base $\{e_1, ..., e_N\}$ of $\mathbf{R}^N$ we can write:

$$D_\epsilon M(\epsilon, t)|_{\epsilon=0} \cdot \tilde{\epsilon} = (D_\epsilon w_a(t) \cdot \tilde{\epsilon}\, \overline{w_a(t)} + w_a(t)\overline{D_\epsilon w_a(t) \cdot \tilde{\epsilon}})_{a=1}^N$$
$$= [2Re(D_\epsilon w_a(t) \cdot \tilde{\epsilon}\, \overline{w_a(t)})]_{a=1}^N$$
$$= [2Re(< D_\epsilon w(t) \cdot \tilde{\epsilon}, e_a > \overline{w_a(t)})]_{a=1}^N$$

Since

$$(e^{\overline{A}(t-s)}\tilde{\epsilon}(s)\overline{B}e^{\overline{A}s})_{ab} = e^{-i\lambda_a(t-s)}\tilde{\epsilon}(s)(-i)B_{ab}e^{-i\lambda_b s} \qquad (17)$$

and taking into account the explicit formula for $w_a(t)$

$$w_a(t) = e^{-i\lambda_a t}w_a(0), \quad a = 1, ..., N \qquad (18)$$

one obtains first

$$D_\epsilon M(\epsilon, t)|_{\epsilon=0} \cdot \tilde{\epsilon} = [2Re(-i\sum_{b=1}^N \int_0^t e^{-i\lambda_a t}B_{ab}e^{i\omega_{ab}s}\tilde{\epsilon}(s)w_a(0)\overline{w_b(t)}ds)]_{a=1}^N$$

and then

$$D_\epsilon M(\epsilon, t)|_{\epsilon=0} \cdot \tilde{\epsilon} = [2Re(-i\sum_{b=1}^N \int_0^t w_b(0)\overline{w_a(0)}B_{ab}e^{-i\omega_{ab}(t-s)}\tilde{\epsilon}(s)ds)]_{a=1}^N \qquad (19)$$

Armed with this formula we are ready to tackle with the local controllability problem. This is in fact a particular surjectivity property of $M(\epsilon, t)$. We will fix $t = T \neq 0$ and will prove that $D_\epsilon M$ has the surjectivity property we desire; by the implicit function theorem the conclusion will follow then for M itself.

We prove that $D_\epsilon M(\epsilon, T)$ is onto the linear manifold (P) (product of hyper-planes of $\mathbf{R}^{cardinality(V_\alpha)}$, $\alpha = 1, .., K$):

$$\{(x_i)_{i=1}^N \in \mathbf{R}^N; \sum_{\{i; \Psi_i \in V_\alpha\}} x_i = 0, \ \alpha = 1, .., K\}$$

whose $M(0, T)$-translation is tangent in $(0, T)$ to the range of M.

Denote by f_a, $a = 1, ..., N$ the components of $D_\epsilon M$:

$$D_\epsilon M(\epsilon, t)|_{\epsilon=0} \cdot \tilde{\epsilon} = (< f_a, \tilde{\epsilon} >_{L^2})_{a=1}^N \qquad (20)$$

Due to the finite dimensionality of our setting we just have to show that the range of $D_\epsilon M(\epsilon, t)|_{\epsilon=0}$ has a null orthogonal with respect to (P), that is any vector $k = (k_a)_{a=1}^N \in \mathbf{R}^N$ such that

$$\sum_{\{i; \Psi_i \in V_\alpha\}} k_i = 0, \quad \alpha = 1, .., K \qquad (21)$$

$$\sum_{i=1}^N k_i \cdot < f_i, \tilde{\epsilon} >_{L^2} = 0, \ \forall \tilde{\epsilon} \in L^2([0, T]) \qquad (22)$$

is necessary the null vector.

The relation (22) can also be written $\sum_{i=1} k_i \cdot f_i(s) = 0$, $\forall 0 \leq s \leq T$ or, in full format,

$$\sum_{a,b=1}^{N} k_a B_{ab} \cdot 2Re[iw_b(0)\overline{w_a(0)}e^{-i\omega_{ab}(T-s)}] \equiv 0, \ \forall 0 \leq s \leq T \tag{23}$$

Grouping together similar terms one gets for all $0 \leq s \leq T$

$$\sum_{a<b}^{N}(k_a - k_b)|w_a(0)||w_b(0)|B_{ab} \cdot 2Re[i\frac{w_b(0)}{|w_b(0)|}\frac{\overline{w_a(0)}}{|w_a(0)|}e^{-i\omega_{ab}(T-s)}] = 0 \tag{24}$$

It suffices now to notice that since in $\{\omega_{ab}; a < b\}$ there are no repetitions, the functions of s are all incommensurable and of null sum. Therefore coefficients are all zero:

$$(k_a - k_b)|w_a(0)||w_b(0)|B_{ab} = 0, \ a,b = 1,...,N, \ a < b \tag{25}$$

Working on connected components of the transfer graph if follows that $k_a = const$, for all a such that $\Psi_a \in V_\alpha$, $\alpha = 1,..,K$ which together with (21) implies that k is the null vector, q.e.d.

Remark 2 *One can view this local controllability theorem as an encouraging argument when designing numerical algorithms. Indeed one is sure that a well designed algorithm will at least be able to improve the initial guess $\epsilon(t) \equiv 0$. It is however the next result that positively informs us about the possibility of solving such a control problem.*

Remark 3 *There is an interesting property of the control solutions that this result highlights: the possibility of **synchronous** control. The theorem states that when working with several independent quantum systems (each modeled by a connected component of the transfer graph G) one can control one of them without interfering with the others (or controlling at the same time the others also); this may give an indication about when control in liquid phase or in other cases of mixtures of systems (one of them being "principal") may be possible: when the systems have different spectral signatures (non-degenerate transitions); this may eventually allow us to choose the right "secondary" systems to accompany our target.*

Remark 4 *The fact that there will always be (at least locally) control solutions that may be chosen to solve (finitely many) other control problems at the same time with our main control problem is suggesting that there is a rich diversity (and hence multiplicity) among the control solutions; therefore for a particular solution $\epsilon(t)$ only a part of all the information contained in $\epsilon(t)$ is useful for reaching the target, all the rest being only some sort of noise. One illustration of how little information ($N - 1$ Fourier coefficients) it may take to reach a population distribution is given in the proof of the next result.*

5.2   Global controllability

Theorem 3 *Under the hypothesis [H] the population distribution of the system (5) is controllable.*

Proof. Let us use the variable substitution $w_{\epsilon k}(t) = e^{i\lambda_k t} c_{\epsilon k}(t)$, $k = 1, .., N$. Then the equations (5) become:

$$\begin{cases} i\frac{\partial}{\partial t} w_{\epsilon k}(t) = \sum_{l \neq k} \epsilon(t) e^{i(\lambda_k - \lambda_l)t} B_{kl} w_{\epsilon l}(t), & k = 1, .., N \\ w_{\epsilon k}(t = 0) = c_{0k}, & k = 1, .., N \end{cases} \tag{26}$$

Since $|w_{\epsilon k}(t)| = |c_{\epsilon k}(t)|$, $\forall t \geq 0$ studying the controllability of (5) is equivalent to studying the controllability of (26). Regarding our definition of the controllability we understand that the goal is in fact to "rearrange" the population distributions inside each connected component of G i.e. to transfer population among vertices belonging to the same connected component.

Let us define an elementary population transfer of μ units between the eigenstate Ψ_k and the eigenstate Ψ_l by the final conditions

$$\begin{cases} |w_{\epsilon k}(T)|^2 = |w_{\epsilon k}(0)|^2 - \mu \\ |w_{\epsilon l}(T)|^2 = |w_{\epsilon l}(0)|^2 + \mu \\ |w_{\epsilon i}(T)|^2 = |w_{\epsilon i}(0)|^2, \ \forall \ i \neq k, l \end{cases} \tag{27}$$

Then it is easy to see that our problem can be decomposed into elementary population transfers between the eigenstates. In fact one can choose these transfers to happen between edges of the graph G^8 i.e. between states Ψ_k and Ψ_l such that $B_{kl} \neq 0$. Of course these population transfers need only be (arbitrary precise but) approximate.

Let us choose $\epsilon(t)$ of the form $\epsilon(t) = \frac{1}{p} r_{kl} cos(\omega_{kl} t) = \frac{1}{p} r_{kl} \frac{e^{i\omega_{kl} t} + e^{-i\omega_{kl} t}}{2}$.
Then (26) can be written in the form:

$$i\frac{\partial}{\partial t} w_{\epsilon a}(t) = \sum_{b \neq a} \epsilon(t) e^{i\omega_{ab} t} B_{ab} w_{\epsilon b}(t), \quad a = 1, .., N, \tag{28}$$

$$i\frac{\partial}{\partial t} w_{\epsilon a}(t) = \sum_{b \neq a} \frac{1}{p} r_{kl} B_{ab} \frac{e^{i(\omega_{ab} + \omega_{kl})t} + e^{i(\omega_{ab} - \omega_{kl})t}}{2} w_{\epsilon b}(t) \quad a = 1, .., N \tag{29}$$

One can see that the only terms that do contain non-oscillatory functions are $i\frac{\partial}{\partial t} w_{\epsilon k}(t)$ and $i\frac{\partial}{\partial t} w_{\epsilon l}(t)$ since in this case one gets in the second term quantities like $\frac{e^{i(\omega_{kl} - \omega_{kl})t}}{2} = \frac{1}{2}$ or $\frac{e^{i(\omega_{lk} + \omega_{kl})t}}{2} = \frac{1}{2}$. We will show that in the limit $p \to \infty^9$ for the

---

[8]Since each G_α is connected it contains at least a tree; one can prove recursively that in fact at most $N - 1$ such operations are needed. Moreover since each G_α remains connected after having eliminated all degenerate transitions one can suppose the transitions correspond to edges which have not been eliminated.

[9] Note that even if the field is taken to be small, we do not enter the classical perturbative framework as our dynamics is still non-linear.

final time pT (T=fixed) all other (oscillatory) terms can be neglected. Indeed let us replace $w_{\epsilon a}(t)$, $a = 1, ..., N$ by $U_a(t)$, $a = 1, ..., N$ given by:

$$\begin{cases} U_k(t) = \frac{w_{\epsilon k}(t) + w_{\epsilon l}(t)}{2} \cdot e^{i\frac{1}{2p}r_{kl}B_{kl}t} \\ U_l(t) = \frac{w_{\epsilon k}(t) - w_{\epsilon l}(t)}{2} \cdot e^{i\frac{1}{2p}r_{kl}B_{kl}t} \\ U_a(t) = w_{\epsilon a}(t), \quad a = 1, ..., N, \quad a \neq k, l \end{cases} \tag{30}$$

Then the evolution system is now

$$i\frac{\partial}{\partial t}U_a(t) = \sum_{b \neq a} \frac{1}{p} r_{kl} B_{ab} f_{ab}(t) U_b(t) \tag{31}$$

where the functions f_{ab} are sums of exponentials e^{iqt} with q having one of the forms $\omega_{ab} - \omega_{kl}$, $\omega_{ab} + \omega_{kl}$, $2\omega_{kl}$, $\omega_{ak} - \omega_{kl} + \frac{1}{2p}r_{kl}B_{kl}$, ... What is important about the frequencies q is that we are able to bound their absolute values by two constants $c_1, c_2 > 0$ that do not depend of p as soon as p is large enough: $0 < c_1 < |q| < c_2 < \infty$.

We will now show that for p large enough $|U_a(pT) - U_a(0)|$, $a = 1, ..., N$ is as small as we want (T is fixed). Let us denote $g_{ab} = r_{kl} B_{ab} f_{ab}(t)$. Denote also by $G_{ab}(t)$ the primitive of g_{ab} that is zero for $t = 0$. From the form of the functions f_{ab} we see that there exists a constant C_0 independent of p such that $|G_{ab}(t)| < C_0, \forall\, t \in \mathbf{R}$. Then

$$iU_a(pT) = iU_a(0) + \int_0^{pT} i\frac{\partial}{\partial t}U_a(t)dt = iU_a(0) + \int_0^{pT} \sum_{b \neq a} \frac{1}{p} g_{ab} U_b(t) dt$$

$$= iU_a(0) + \sum_{b \neq a} \frac{1}{p} G_{ab}(pT) U_b(pT) + i\int_0^{pT} \sum_{b \neq a} \frac{1}{p} G_{ab}(t) i\frac{\partial}{\partial t}U_b(t) dt$$

$$= iU_a(0) + \sum_{b \neq a} \frac{1}{p} G_{ab}(pT) U_b(pT) + \frac{i}{p}\int_0^{pT} \sum_{b,c=1}^{N} G_{ab}(t) \frac{1}{p} g_{bc} U_c(t) dt$$

Using the fact that G_{ab}, g_{ab} and $U_a(t)$ are bounded functions on $\mathbf{R}$ it follows that for each $\eta > 0$ we can choose p large enough such that $|iU_a(pT) - iU_a(0)| < \eta$. After having replaced back the U_a in the system we conclude that for p large enough the solutions of the system (29) computed in $t = pT$ are as close as we want to the solutions of

$$\begin{cases} i\frac{\partial}{\partial t}w_{\epsilon k}(t) = \epsilon(t)e^{i\omega_{kl}t}B_{kl}w_{\epsilon l} \\ i\frac{\partial}{\partial t}w_{\epsilon l}(t) = \epsilon(t)e^{i\omega_{lk}t}B_{kl}w_{\epsilon k} \\ i\frac{\partial}{\partial t}w_{\epsilon a}(t) = 0, \quad a = 1, ..., N, \quad a \neq k, l \end{cases} \tag{32}$$

A straightforward analysis of the case $N = 2$ proves now that one can realize any desired population transfer by tuning the coefficient r_{kl} depending on how many "population units" are to be transfered between the eigenstates Ψ_k and Ψ_l.

Remark 5 *The hypothesis (H) is verified in a large class of practical cases (see [13])*[10]. *Moreover there are examples where the absence of this hypothesis makes the system not controllable. One can consider for instance the case of a system made up by two identical and independent sub-systems. It is obvious that using the same laser pulse one cannot obtain different results for the components. The hypothesis (H) is here to prevent such correlations* **induced by the similarities in the spectral signatures** *to go unnoticed.*

Remark 6 *Even if our approach is constructive it is not entirely optimal; one can see that there are simple ways to reduce the time required to reach the target by constructing simultaneously elementary transfers. In order to formalize this one should optimize a distributed transport problem*[11] *on the graph* G. *On the other side numerical results suggest us to conjecture that the* L^2 *norm of the field* $\epsilon(t)$ *realizing the transfer remains constant.*

As an improvement of the result above one can prove the following

Theorem 4 *Let* d_0 *be the population distribution associated to the initial state* C_0: $d_0 = (|c_{0i}|)_{i=1,...,N}$. *Under the hypothesis [H] any population distribution* $d = (d_i)_{i=1}^N$ *such that* $d_i > 0$, $i = 1, ..., N$ *and that verifies the necessary conditions (12) can be* **exactly** *reached.*

Proof. The proof combines the global approximate controllability with a slightly stronger form of the local controllability result. In order to simplify the presentation let us denote for any state C by $d(C)$ its associated population distribution: $d(C) = (|C_i|)_{i=1}^N$. We will first assume the following

Lemma 2 *Let* C_l *be an initial state and* $d_l = d(C_l)$. *Suppose* $d_{li} > 0$, $i = 1, ..., N$ *and that the hypothesis [H] is verified. Then there exists an open neighborhood* D_l *of* d_l *(on the surface of* $\mathbf{R}^N$ *given by the necessary conditions (12) endowed with the canonical topology) such that any population distribution* d *in* D_l *can be exactly reached from from any initial state* C_i *with* $d(C_i) \in D_l$.

In order to prove the theorem one applies the lemma above for d and obtains an open neighborhood D. By the global result there exists a field that drives the system from the initial state C_0 to some state C_1 with $d(C_1) \in D$. Since d is obviously in any of its open neighborhoods (and in particular in D) it follows that there exists a field that drives C_1 to d. All that remains to be done is to "glue" those two parts together, obtaining thus an field that allows to exactly reach d starting from C_0 (and passing by d_1).

Let us now prove the lemma (2). Note first that the local controllability result is uniform with respect to the initial state. Indeed, with the same notations as in the proof of theorem (2), note that $c(\epsilon, C, T)$ is of C^1 class with respect to C. Moreover we have proved that the (Fréchet) derivative of $(\epsilon, C, T) \rightarrow (|c_a(\epsilon, C, t)|^2)_{a=1}^N$ with

---

[10]When this theory cannot be used some positive results may still be available; it is the case of the harmonical oscillator, see [9]; let us note however the need for a particular perturbation regime.

[11]see [5]

respect to ϵ computed at $(0, C, T)$ is surjective. This allows us to apply the implicit function theorem and conclude that there exists a neighborhood D_C of C and a neighborhood (for the topology cited above) D of d such that from any initial state $C_i \in D_C$ one can exactly reach any population distribution in D.

Suppose now that the lemma (2) is not true. Then there exists a sequence of states C_n (with $d_n = d(C_n)$) and a sequence of distributions d_n^t such that $d_n \to d$, $d_n^t \to d$ and d_n^t cannot be exactly reached from C_n. Each C_n is characterized by the set of phase factors (that can be safely supposed to be in $[0, 2\pi]^N$) and population distribution d_n. Since the set $[0, 2\pi]^N$ is compact one can extract from the sequence of phase factors one sub-sequence that is converging; toghether with $d_n \to d$ it follows that one can find a sub-sequence of C_n that is converging to some state C with $d(C) = d$. In conclusion there exists a sequence $C_{n_k} \to C(k \to \infty)$ of states and a sequence $d_{n_k}^t \to d$ of population distributions such that one cannot exactly reach $d_{n_k}^t$ from C_{n_k} for any $k \geq 1$. This is obviously contradicting the uniform properties of the local result, q.e.d.

6   Application to a five level system

As an application of the results above we will study a five-level system presented in [17]. It will be seen that the controllability is easy to check by our method; we will take advantage of the constructive side of the theory to support the theoretical results in numerical simulations.

The matrix representation of the operators involved are:

$$
A = \begin{pmatrix} 1.0 & 0 & 0 & 0 & 0 \\ 0 & 1.2 & 0 & 0 & 0 \\ 0 & 0 & 1.3 & 0 & 0 \\ 0 & 0 & 0 & 2.0 & 0 \\ 0 & 0 & 0 & 0 & 2.15 \end{pmatrix}, \; B = \begin{pmatrix} 0 & 0 & 0 & 1 & 1 \\ 0 & 0 & 0 & 1 & 1 \\ 0 & 0 & 0 & 1 & 1 \\ 1 & 1 & 1 & 0 & 0 \\ 1 & 1 & 1 & 0 & 0 \end{pmatrix} \tag{33}
$$

One can deduce the following transfer graph G:

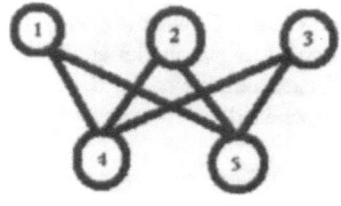

$$\tag{34}$$

Since G is obviously connected and since the hypothesis (H) is also verified it follows that the population distribution of this system is controllable. We present some numerical examples of population transfer, drawing in each case the evolution of the populations $(|c_i(t)|^2, \; i = 1, .., 5)$ in the corresponding viewgraphs.

In the first case we want to transfer population from state 1 directly to state 5. We write this control problem formally $(1, 0, 0, 0, 0) \rightarrow (0, 0, 0, 0, 1)$. Using a laser field of the form $\epsilon(t) = \beta cos((\lambda_1 - \lambda_5)t)$ one obtains the following population evolution [9]:

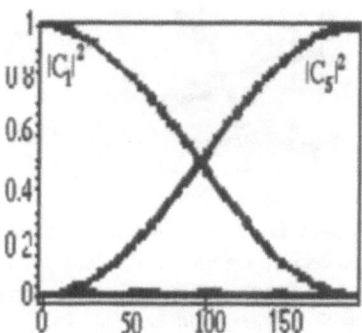

In the second example the target is the state 2 and we use state 4 as intermediary: $(1, 0, 0, 0, 0) \rightarrow (0, 1, 0, 0, 0)$. The laser takes now the form $\epsilon(t) = \beta_1 cos((\lambda_1 - \lambda_4)t) + \beta_2 cos((\lambda_4 - \lambda_2)t)$.

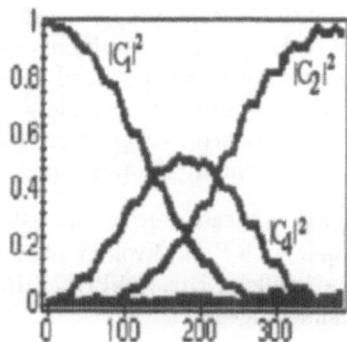

The last case is a cooperative control example where population is flowing to target state 4 from states 1 and 2: $(\frac{1}{3}, \frac{2}{3}, 0, 0, 0) \rightarrow (0, 0, 0, 1, 0)$. The laser has the same general form.

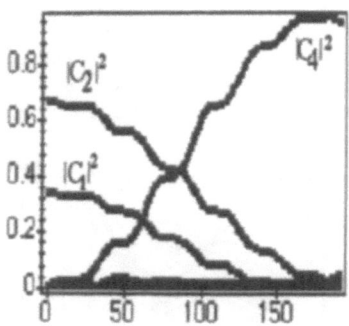

Remark 7 *The transfer speed is essentially given by the matrix B. Studying the time needed for the three experiments one notes that for controls of the same order in L^∞ norm the second is two times slower since the population has to go from Ψ_1 to Ψ_4 and then from Ψ_4 to Ψ_2; that could be also realized by two simulation of the first type.*

Remark 8 *With values in $O(1)$ in the B matrix in order to obtain a precision of 10^{-2} our method requires heuristically a "p" of order $O(10^2)$ which is consistent (and of the same order of magnitude) with the time usually obtained in the literature (see [17]).*

7 Conclusions

Controllability of the bilinear quantum systems has been studied in the infinite and finite dimensional settings. The infinite dimensional case has been seen to exhibit hidden restrictions due to some compacity properties of the equations involved. For the finite dimensional case and as long as one is interested in the populations of the eigenstates, positive results have been obtained for exact local and for global controllability. Easy to check and intuitively simple to understand necessary and sufficient conditions have been obtained to characterize the attainable set. Numerical simulations for a five level system are also presented.

Acknowledgements. It is a pleasure to acknowledge helpful informal discussions that we had on this topic with Prof. Yvon Maday (ASCI Laboratory). Useful references and remarks from Claude Le Bris (ENPC-CERMICS) and Pierre Rouchon (CAS-ENSMP) are also acknowledged.

8 Appendix

Even if the physical interpretation for $|c_i(t)|^2, i = 1, ..., N$ as populations is valid only in the case of a diagonal hamiltonian our results can be extended for general auto-adjoint internal hamiltonian matrix H_0 and dipole operator B.

Indeed since H_0 is auto-adjoint there exists a $(N \times N)$ unitary transform U such that $U H_0 U^t$ be diagonal. It is then straightforward to state the above results for the system so transformed.

What may be more interesting to notice is that B_{ii} need not be zero in order for the controllability results to remain true. This is easy to prove for the local controllability result: the set of formulae (25) remains the same since for $a = b$ the term in the formula (23) becomes:

$$k_a B_{aa} 2Re[iw_a(0)\overline{w_a(0)}e^{-i\omega_{aa}s}] = k_a B_{aa} 2Re[i|w_a(0)|^2] = 0$$

For the global controllability result one has to work with a modified variable substitution $w_{\epsilon k}(t) = e^{i\,(l_k + \epsilon(t)B_{kk})t}c_{\epsilon k}(t)$, $k = 1,..,N$. Since the L^∞ norm of $\epsilon(t)$ is to be chosen small enough later in the proof, one realizes that in fact the uniform (in p and t) boundeness of the $g_{ab}(t), G_{ab}(t)$ still holds true, which is enough to obtain the conclusion of the theorem.

Bibliography

[1] J.M.Ball, J.E.Madersen and M.Slemrod, "Controllability for distributed bilinear systems", SIAM J.Control and Optimization, vol 20 (4) (1982), 575–597

[2] C. Le Bris, "Control Theory Applied to Quantum Chemistry: Some Tracks" ESAIM PROC, to appear

[3] P.Brumer and M.Shapiro, "Coherence Chemistry: Controlling Chemical Reactions with Lasers", Acc.Chem Res. 22, 12 (1989) 407–413.

[4] A.G. Butkovskiy, Yu.I.Samoilenko, "Control of quantum-mechanical processes and systems", Kluwer,1990

[5] C.W.Churchmann, L.Ackoff, E.Arnoff, "Introduction à la recherche opérationnelle", Dunod, Paris, 1970

[6] M.Demilrap and H.Rabitz, Phys. Rev. A., **47** 2 1983, p.831

[7] C.M. Dion et al., Chem. Phys.Lett 302(1999), 215-223

[8] C.M. Dion, A.Keller, O.Atabek & A.D. Bandrauk, Phys. Rev. A 59(2) 1999, p.1382

[9] Kime K., "Control of transition probabilities of the quantum-mechanical harmonic oscillator", Appl. Math. Lett. 6 (3) (1993) 11–15.

[10] Huang G.M., Tarn T.J., Clark J.W., "On the controllability of quantum-mechanical systems", J. Math. Phys. 24, 11 (1983) 2608–2618.

[11] Mei Kobayashi,"Mathematics make molecules dance" , SIAM News 24 (1998)

[12] A.P.Pierce, M.A. Dahleh and H.Rabitz, Phys Rev.A **37** (1998), p.4950

[13] Ramakrishna V. & al., "Controlability of molecular systems" Phys. Rev. A 51 (2) (1995) 960–966.

[14] Shi S., Rabitz H., "Optimal control of selectivity of unimolecular reactions via an excited electronic state with designed lasers" , Chem. Phys. 97 (1992) 276–287.

[15] S.Shi, A.Woody, and H.Rabitz, "Optimal control of selective vibrational excitation in harmonic linear chain molecules" , J.Chem Phys. 88(1988), p.6870

[16] Tannor D.J., Rice S.A., "Control of selectivity of chemical reaction via control of wave packet evolution", J. Chem. Phys. 83 (1985) 5013–5018.

[17] S.H. Tersigni, P.Gaspard and S.A. Rice, "On using shaped light pulses to control the selectivity of product formation in a chemical reaction: An application to a multiple level system", J. Chem. Phys. 93, 3(1990) 1670–1680.

[18] G. Turinici, "Analysis of Numerical Simulation Methods in Quantum Chemistry", Ph.D. Thesis, work in progress

[19] W.S.Warren, H.Rabitz and M.Dahleh, Warren W.S., Rabitz H., Dahleh M., "Coherent control of quantum dynamics : the dream is alive", Science 259 (1993) 1581–1589.

Part II

Condensed phases

Chapter 5

Recent mathematical results on the quantum modeling of crystals

I. Catto [1], **C. Le Bris** [2] **and P.-L. Lions**[1]
[1] *CEREMADE (UMR CNRS 7534),*
Université Paris-Dauphine,
Place du Maréchal de Lattre de Tassigny,
F-75775 Paris Cedex 16, France.
`{catto,lions}@ceremade.dauphine.fr`
[2] *CERMICS, Ecole Nationale des Ponts et Chaussées,*
6 & 8, avenue Blaise Pascal, Cité Descartes,
F-77455 Marne-La-Vallée Cedex, France
`lebris@cermics.enpc.fr`

Abstract : We describe in this paper a strategy, the so-called thermodynamic limit process, to build in a rigorous mathematical manner the quantum-mechanical models for the ground-state energy of solid crystals. These models are the analogues for the solid state of well-known models issued from Quantum Chemistry, namely Thomas-Fermi, Hartree and Hartree-Fock type models. We shall present a broad overview on recent mathematical studies on this topic.

1   Introduction

Many mathematical studies have been devoted to the thermodynamic limit problem as stated in its full generality within the Statistical Mechanics, and we refer in particular the reader to the works by Ruelle [52], Lebowitz and Lieb [35, 39, 40], and Fefferman [28, 29]. We shall not deal here with the physical background of this theoretical problem, and we refer the reader to the textbooks [7, 59] and the articles [36, 40]. The works we shall present here are focused on models for charged particles at the zero temperature limit, where in some sense, they are frozen at their ground-state, and even more specifically on the crystalline phase.

Briefly speaking, the so-called thermodynamic limit problem consists in examining the behaviour of models for particles in a finite volume of matter when the volume under consideration goes to infinity. Since the energy is an extensive thermodynamic quantity, it is expected that the energy per unit volume goes to a finite limit when the volume goes to infinity. It is also expected that the function representing the state of the matter goes also to a limit in some sense. The thermodynamic limit problem we study (that is, for crystals, and at zero temperature) may be stated as follows.

We consider a neutral molecular system consisting of N electrons and N nuclei of unit charge (atomic units will be adopted in all that follows). According to the Born-Oppenheimer approximation, the electronic ground-state is first determined for a given nuclear configuration. The geometry of the nuclei and their electric charge will be taken as fixed and arranged according to a crystalline lattice. To fix ideas, from now on, our reference lattice will be the lattice of points of integer coordinates—that is, $\mathbf{Z}^3$—and therefore the primitive cell of the crystal will be the unit cube, denoted by Q in the sequel. The N point nuclei of unit charge being located on a finite subset Λ_N of $\mathbf{Z}^3$ (each point being at the center of a cubic unit cell), we asymptotically let the set Λ_N fill in the entire lattice $\mathbf{Z}^3$. The union of all cubic cells whose center is a point of Λ_N is denoted by $\Gamma(\Lambda_N)$; it has volume N. It is important to note that, in all that follows, $\Gamma(\Lambda_N)$ may be viewed as a big box to which the molecule is confined. (This claim may actually be checked in a rigorous mathematical way.) This assumption is standard for statistical physicists, and is compulsory at positive temperature. It turns out that, at zero temperature, any boundary conditions for the wave functions or the electronic density on a big box one may think of (like Neumann, Dirichlet or periodic boundary conditions) gives rise to the same periodic model after passing to the thermodynamic limit. The boundary conditions may therefore be chosen conveniently according to everyone's own preferences, for numerical purposes, for instance. Let us simply mention that Solid State physicists seem to prefer to work with the so-called Born-Von Karman conditions, which are simply periodic boundary conditions on the boundary of a big parallelepiped box.

Suppose that for any subset of $\mathbf{Z}^3$ fixed, we have a well-posed model for the ground-state of the neutral molecule consisting of N electrons and N nuclei located at the points of Λ_N. Let us denote by I_{Λ_N} the ground-state energy, and by ρ_{Λ_N} the minimizing electronic density. Then, the question of the existence of the thermodynamic limit for the model under consideration may be stated as follows :

(i) Does there exist a limit for the energy per unit volume $\frac{1}{N}I_{\Lambda_N}$ when N goes to infinity ?

(ii) Does the minimizing electronic density ρ_{Λ_N} approach a limit ρ_{∞} (in a sense to be made precise later) when N goes to infinity ?

(iii) Does the limit density ρ_{∞} have the same periodicity as the assumed periodicity of the nuclei ?

We investigate the behaviour of the ground-state energy per unit volume, as N goes to infinity, within standard approximation theories in Quantum Chemistry for the ground-state energy; namely, models from Density Functional Theory, like Thomas-Fermi type models on the one hand, and the restricted Hartree, the Hartree, the reduced Hartree-Fock and the Hartree-Fock models, on the other hand. Let us emphasize the fact that in all that follows we shall not deal with more advanced models from Density Functional Theory like the Kohn-Sham or the $X\alpha$ model (see e. g. [25]), since from a mere mathematical viewpoint this kind of models features the same structure, and thereby the same difficulties, as the Hartree-Fock model. In particular, the results that we describe within the setting of the Hartree-Fock model carry through just as they are to a wide class of models from modern Density Functional Theory.

By applying this strategy to various models, we have different purposes in mind. Firstly, we want to check that the molecular model under consideration does have the good behaviour in the limit of large volumes, thereby validating it a bit further for large molecules, or, on the contrary, limiting its efficiency to modeling true phenomena. Secondly, bearing in mind that our purpose is to build new models for the solid phase, we wish to set a limit problem that is well-posed mathematically and that can be justified in the most rigorous way. With this model at our disposal, we intend to give a sound ground to numerical simulations of the condensed phase. Thirdly, this method aims at linking non-linear quantum-mechanical descriptions of crystals to concepts and techniques issued from Solid State Physics. A further step beyond is next to make use of this models to calculate macroscopic quantities, such as elasticity constants, for example.

Let us first summarize the results which are available on these issues, as far as we know. It has been proved by Lieb and Simon [41] for the Thomas-Fermi model, and by the authors in [16] for the Thomas-Fermi-von Weizsäcker model (which is a variant of the Thomas-Fermi model—see below) that the ground-state energy per unit volume converges to a periodic minimization problem, which is set on the unit cell of the crystal, and that the electronic density also converges to a periodic electronic density, which is precisely the minimizer of the limit periodic minimization problem.

The situation is far more complicated for more realistic models which involve the electronic wave-functions for describing the electronic state, as the Hartree and the Hartree-Fock models. One obvious reason is that we have to analyze the behaviour of an increasing number N of wave-functions and not only of one function, the electronic density. For simplified forms of the Hartree and the Hartree-Fock models, the restricted Hartree and the reduced Hartree-Fock models respectively, we completely determine the asymptotic behaviour, proving that the limit is actually a minimization problem set on the unit cell of the lattice. However, we have only extremely weak results on the behaviour of the electronic density to a periodic density, which are not very satisfactory.

For the Hartree and the Hartree-Fock models, we are only able to define periodic problems that are likely to be the thermodynamic limits. Within these models, the mere definition of the limit problem turns out to be a substantial piece of the work (writing a periodic problem that has some rigorous mathematical sense is not straightforward at all). In order to prepare and stimulate future works on the subject, we prove that the periodic limit problems that we define are well-posed. In particular, they allow to derive Hartree and Hartree-Fock type equations for crystals, which may be found in Solid State physics textbooks, thereby validating our method. Let us finally mention that, in order to set the periodic models which are the analogues for the crystalline phase of the molecular version of the Hartree-Fock and the reduced Hartree-Fock models, we extensively rely upon the decomposition of operators commuting with the group of translations of $\mathbf{Z}^3$. Consequently, we observe that the theory of Bloch waves, which is commonly used by Solid State physicists, appears naturally in the nonlinear framework we are considering.

This paper is organized as follows. In the following section, we state the thermodynamic limit problem within the framework of Density Functional Theory, and more precisely, in the Thomas-Fermi setting (and related more advanced models). We are aware that these models are neither the most efficient ones nor the ones which are usually used for calculations. Nevertheless, very fundamental (and difficult) issues are already raised within these models, which one has also to face while dealing with more relevant models, like Hartree-Fock for example. By testing first our method on the Thomas-Fermi type models, we actually try to overcome the technical difficulties step by step as the complexity (together with the relevance) of the model increases. Section 3 is devoted to Hartree type models, while Hartree-Fock type models are treated in Section 4.

In Section 5, we shall report on recent progress made by Blanc and one of us [9, 11, 12] on the existence of a minimal energy configuration of the nuclei inside the class of all possible periodic lattices for the TFW model.

Finally, in Section 6, we suggest different extensions of all these works.

2   Models from Density Functional Theory

This section is devoted to the well-known Thomas-Fermi model and some of its variants. These models belong to a large class of models that are today identified as the models arising in Density Functional Theory. We refer the reader to [25, 46] for an introduction to the general features and the physical foundations of such models, and to [14, 21, 22] for a comprehensive list of recent mathematical results on all the models which are considered throughout this paper.

Within the Thomas-Fermi type models, the density functional may be written as follows :

$$E_{\Lambda_N}^{TFW}(\rho) = c_0 \int_{\mathbf{R}^3} |\nabla \sqrt{\rho}|^2 + c_1 \int_{\mathbf{R}^3} \rho^{5/3} - \sum_{k \in \Lambda_N} \int_{\mathbf{R}^3} \frac{\rho(x)}{|x - k|} \, dx$$
$$+ \frac{1}{2} \iint_{\mathbf{R}^3 \times \mathbf{R}^3} \frac{\rho(x)\rho(y)}{|x - y|} \, dx dy + \frac{1}{2} \sum_{y \neq z \in \Lambda_N} \frac{1}{|y - z|}, \qquad (1)$$

and the ground-state energy is given by

$$I_{\Lambda_N}^{TFW} = \inf \left\{ E_{\Lambda_N}^{TFW}(\rho); \rho \geq 0, \ \sqrt{\rho} \in H^1(\mathbf{R}^3), \int_{\mathbf{R}^3} \rho = N \right\}. \qquad (2)$$

In (1) above, c_0 and c_1 are non-negative constants (whose exact values are irrelevant from a mathematical view-point). The Thomas-Fermi model (TF, for short) corresponds to the case when $c_0 = 0$, while the full model is referred to as the Thomas-Fermi-von Weizsäcker model (TFW, for short). In all that follows, we shall make the convention that $c_0 = c_1 = 1$ in the TFW setting. It is to be noticed, that in both cases, the energy functional is strictly convex with respect to the electronic density.

Many mathematical studies have been devoted to these models, and, in particular, it is a well-known fact that the problem (1)-(2) has a unique minimizing density, denoted by ρ_{Λ_N}. This fact has been proved by Lieb and Simon [41] (see also [37]) for the Thomas-Fermi model, and by Benguria, Brézis and Lieb [8] (see also [37]) for the TFW model.

Before turning to the thermodynamic limit problem *per se*, let us make some comments. The TFW model remedies major non-physical features of the Thomas-Fermi model. First of all, it predicts the expected behaviour of the density near the nuclei (*cusp conditions*) and at large distance (*exponential fall-off of the electronic density*). Moreover, as shown by two of us in [20], binding of neutral molecules holds in this model, at the difference with Teller's no binding theorem for Thomas-Fermi model.

We have on purpose omitted to mention until now the ground-breaking work [41] by Lieb and Simon on the thermodynamic limit in the framework of the Thomas-Fermi theory. Indeed, this work is at the origin of our own study [16] on the Thomas-Fermi-von Weizsäcker model, and has therefore a far larger impact on our work than the, however fundamental, works that we have quoted in the introduction. Lieb and Simon proved in [41] that the three questions **(i)-(ii)-(iii)** of the thermodynamic limit problem that we have addressed in the introduction can be answered positively in the setting of the TF theory. In particular, they set a Thomas-Fermi type model for crystals. It is to be mentioned however that their strategy of proof relies very much upon Teller's no binding theorem, and therefore, it cannot be carried through the TFW setting. In [16], we have proved that the three questions **(i)-(ii)-(iii)** of the thermodynamic limit problem can also be answered positively in the setting of the TFW theory. In addition, the new strategy of proof which we apply in [16] actually allows to extend the results of Lieb and Simon on the Thomas-Fermi model. Moreover, some of the techniques developed there also carry through the Hartree and the Hartree-Fock model.

Let us now introduce the periodic minimization problems obtained by passing to the thermodynamic limit. We then define the following periodic minimization problem set on the unit cell Q of the lattice

$$I_{per}^{TFW} = \inf\{E_{per}^{TFW}(\rho); \rho \geq 0, \sqrt{\rho} \in H_{per}^1(Q), \int_Q \rho = 1\}, \tag{3}$$

$$
\begin{aligned}
E_{per}^{TFW}(\rho) &= \int_Q |\nabla\sqrt{\rho}|^2 + \int_Q \rho^{5/3} - \int_Q \rho(x)G(x)dx \\
&+ \frac{1}{2}\iint_{Q \times Q} \rho(x)\rho(y)G(x-y)dxdy,
\end{aligned} \tag{4}
$$

where $H_{per}^1(Q)$ is the subset of $H_{loc}^1(\mathbf{R}^3)$ consisting of functions which satisfy the periodic boundary conditions on the boundary of Q. The potential G which appears in the definition (4) of the TFW functional is defined, in a unique way, by

$$-\Delta G = 4\pi \left(-1 + \sum_{y \in \mathbf{Z}^3} \delta(\cdot - y)\right), \tag{5}$$

and

$$\int_Q G = 0. \tag{6}$$

The main results we obtain in [16] (which were announced in [15]) may be stated as follows. (We need technical assumptions that we do not make precise here.) We first prove the **convergence of the energy per unit volume**:

$$\lim_{N \to +\infty} \frac{1}{N} I_{\Lambda_N}^{TFW} = I_{per}^{TFW} + \frac{M}{2},$$

where $\frac{M}{2}$ is a universal constant, that only depends on G, through $M = \lim_{x \to 0}[G(x) - \frac{1}{|x|}]$, and which is just a matter of normalization. In particular, the ground state energy behaves linearly with respect to the total volume of the large molecule, and we have rigorously obtained a model for the ground-state energy of the crystal. Independently of this first result, we also establish in [16] the **convergence of the electronic density** in the following sense. The electronic density ρ_{Λ_N} converges to ρ_{per}, the unique periodic minimizing density of I_{per}^{TFW}, uniformly on sequences of domains approaching $\mathbf{R}^3$ in a sense defined precisely in [16]. Roughly speaking, these domains, which are included in the "big box" $\Gamma(\Lambda_N)$, fill in asymptotically the whole space; their complementary subset in $\Gamma(\Lambda_N)$ has a volume which is small compared to N, and they stay "very far away" from the boundary of the big box. We shall only briefly mention at this stage that, in order to avoid surface effects, technical assumptions are required in [16] on the sequence Λ_N of admissible finite subsets of the lattice. These technical assumptions are well-identified and commonly used in Statistical Physics, and are satisfied by "reasonable big boxes", like cubes, for example.

The above result on the asymptotic behaviour of the ground-state electronic density, when the molecule becomes larger and larger, roughly says that, while the nuclei spread over the whole lattice $\mathbf{Z}^3$, the electronic density becomes periodic, leaving asymptotically one electron per unit cell.

Let us now make some comments on the periodic TFW energy by itself. First of all, it mimics very well the usual TFW energy (1)–(2), except that the integrals involved in (4) are set on the unit cell and that the electronic density shares the same periodicity as the lattice of nuclei. It is also to be remarked that in (4), the periodic potential G which features the electrostatic interaction potential created by the lattice of nuclei models at the same time the self-interaction between the electrons. It is to be noticed that a key-point for the definition of the periodic problem is the definition of laws of interaction between particles, i.e. of the interaction potential(s). Indeed, owing to the long-range of the Coulomb potential, the electrostatic potential created by the infinite lattice of nuclei cannot be simply $\sum_{k \in \mathbf{Z}^3} \frac{1}{|x-k|}$, since this series obviously does not make sense. Besides, each of the three terms appearing in the electrostatic contributions to the ground-state energy (1), namely the *electrons-nuclei interaction* $-\sum_{k \in \Lambda_N} \int_{\mathbf{R}^3} \frac{1}{|x-k|} \rho(x) dx$, the *nuclei-nuclei interaction* $\sum_{y \neq z \in \Lambda_N} \frac{1}{|y-z|}$, and *the electrons-electrons interaction* $\frac{1}{2} \iint_{\mathbf{R}^3 \times \mathbf{R}^3} \frac{\rho(x)\,\rho(y)}{|x-y|}\, dx dy$, is of the order of $N^{5/3}$ as the volume of the box N (which is also the number of particles) goes to infinity. Still, their sum **globally** behaves linearly with respect to N. The compensating effects are explained by the so-called screening phenomenon, once again commonly playing this role in the thermodynamic limit issues—see, for instance, [35, 39, 40]. Let us noteworthy observe, still on this point, that the periodic potential G which was previously defined by (5) and (6) is also

$$G(x) = \sum_{k \in \mathbf{Z}^3} \left(\frac{1}{|x-k|} - \int_Q \frac{1}{|x-y-k|}\, dy \right)$$

(up to a constant); that is the sum over the lattice points of the Coulomb potential created by a point charge placed at the center of the unit cube, and which is screened, on each cell, by a uniform background of negative unit charge. Different techniques are proposed in Solid State physics to compute efficiently such electrostatic sums, like Ewald summation techniques (see, for example, [24, 27], and also [10]).

All these comments on the potential G and the screening effect already appear in Lieb and Simon's work on the Thomas-Fermi model [41]. In particular, in the TFW setting, the potential G is the same as the one appearing in the TF setting, and the periodic minimization problem is rather easy to guess in view of the one arising for the TF theory (they only differ from the gradient term). Likewise, it is easy to check that this periodic minimization problem is mathematically well-posed. In other words, taking benefit from the work by Lieb and Simon who had already defined the TF periodic problem, the idea to introduce the periodic problem (3)–(4) was straightforward. In [16], our "only" contribution was therefore to prove that the TFW model does converge in the thermodynamic limit to (3)–(4). However, concerning the Hartree and the Hartree-Fock models, the situation is much more difficult, owing in particular to the fact that the corresponding energy functional have no more a convex underlying structure, which was of crucial importance in the techniques employed in the TF and the TFW settings. In particular, for these models, guessing the limit periodic model is *per se* a fundamental and difficult step. A (small) part of the work has already been done within the framework of the Thomas-Fermi type models; in particular, one may expect that the limit of the electrostatic contribution to the ground-state energy should be the same; indeed, in every model which is dealt with in the sequel, the energy functional will contain, possibly among other electrostatic terms, the same expression for the electrostatic energy, that is

$$E_{\Lambda_N}^{elec}(\rho) = - \sum_{k \in \Lambda_N} \int_{\mathbf{R}^3} \frac{\rho(x)}{|x - k|} \, dx + \frac{1}{2} \iint_{\mathbf{R}^3 \times \mathbf{R}^3} \frac{\rho(x) \, \rho(y)}{|x - y|} \, dx dy$$

$$+ \frac{1}{2} \sum_{y \neq z \in \Lambda_N} \frac{1}{|y - z|}. \tag{7}$$

Therefore, we may expect that in all the models we shall consider, we shall always have, for a minimizing electronic density ρ_{Λ_N}

$$\lim_{N \to +\infty} \frac{1}{N} E_{\Lambda_N}^{elec}(\rho_{\Lambda_N})$$

$$= -\int_Q G \rho + \frac{1}{2} \iint_{Q \times Q} \rho(x) \rho(y) G(x - y) dx dy + \frac{M}{2}, \tag{8}$$

where ρ is the limit of ρ_{Λ_N}, and is Q-periodic. In spite of the fact that we have not been able until now to prove the thermodynamic limit for these models, we dispose of indications which help us to define periodic minimization problems, which are likely to be the crystalline analogues of the usual Hartree (H for short) and Hartree-Fock (HF for short) models in Molecular Chemistry. We prove that the periodic limit problems that we define are well-posed: There exist minimizers and one may therefore derive meaningful Euler-Lagrange equations. These models will be presented in Sections 3 and 4 respectively; they have been studied by the authors in [18] and [19] respectively, and, for both models, the results have been announced in [17]. Nevertheless, as a preliminary stage to the modeling of the crystalline phase, we have also considered in [18] and [19] (see also [17]), simplified forms of the Hartree (respectively the Hartree-Fock) model, which are referred to as the Restricted Hartree (RH for short) (respectively Reduced Hartree-Fock (RHF for short)) model for which we are able to identify the limit of the energy per unit volume.

Let us begin with the Hartree model.

3   Hartree type models

Let us recall the Hartree model, historically introduced by Hartree in [32]. This model is obtained, when one does not take into account the Pauli exclusion principle for the electrons, and, then simply represents the electronic wave-function as a product of independent wave functions (atomic orbitals). The corresponding ground-state energy is then given by

$$
\begin{aligned}
I_{\Lambda_N}^H \;=\; & \inf \left\{ E_{\Lambda_N}^H(\varphi_1; \cdots; \varphi_N) \; ; \right. \\
& \left. \varphi_i \in H^1(\mathbf{R}^3), \int_{\mathbf{R}^3} \varphi_i^2 = 1, 1 \leq i \leq N \right\},
\end{aligned}
$$

where the φ_i's are the atomic orbitals, and where

$$
E_{\Lambda_N}^H(\varphi_1; \cdots; \varphi_N) = \sum_{i=1}^{N} \left(\int_{\mathbf{R}^3} |\nabla \varphi_i|^2 - \frac{1}{2} D(\varphi_i^2, \varphi_i^2) \right) + E_{\Lambda_N}^{elec}(\rho)
$$

with $\rho = \sum_{i=1}^N |\varphi_i|^2$, and $D(f, g) = \iint_{\mathbf{R}^3 \times \mathbf{R}^3} \frac{f(x)\, g(y)}{|x-y|}\, dx dy$. The existence of (at least) one minimizer for this problem has been established by Lieb and Simon [42], and the existence of infinitely many solutions to the corresponding Euler-Lagrange equations (in other words, infinitely many bound states) is due to Lions [43].

It is not difficult to check that the energy per unit volume is bounded (and, actually, its limit exists), and this implies (see [18]) that the sum of the self-interactions of the electrons, that is $\sum_{i=1}^N D(\varphi_i^2, \varphi_i^2)$, is of the order of N. Therefore, this electrostatic term features a different behaviour, compared to the other terms in $E_{\Lambda_N}^{elec}(\rho)$ (which are of the order of $N^{5/3}$). Hence, one guess could be that this term does not affect the thermodynamic limit. The Restricted Hartree model is the model one obtains by getting rid of this contribution. This model may therefore be written as

$$
I_{\Lambda_N}^{RH} = \inf \left\{ E_{\Lambda_N}^{RH}(\varphi_1; \cdots; \varphi_N) \; ; \varphi_i \in H^1(\mathbf{R}^3), \int_{\mathbf{R}^3} \varphi_i^2 = 1, 1 \leq i \leq N \right\},
$$

$$
E_{\Lambda_N}^{RH}(\varphi_1; \cdots; \varphi_N) = \sum_{i=1}^{N} \int_{\mathbf{R}^3} |\nabla \varphi_i|^2 + E_{\Lambda_N}^{elec}(\rho).
$$

It turns out that the unique minimizer of $E_{\Lambda_N}^{RH}(\varphi_1; \cdots; \varphi_N)$ is of the form $(\frac{\sqrt{\rho_{\Lambda_N}}}{N}; \cdot; \frac{\sqrt{\rho_{\Lambda_N}}}{N})$. And, the limit of the energy per unit volume may be identified as

$$
I_{per}^{RH} = \inf \left\{ E_{per}^{RH}(\rho) \; ; \rho \geq 0, \sqrt{\rho} \in H_{per}^1(Q) \int_Q \rho = 1 \right\} + \frac{M}{2},
$$

where

$$
E_{per}^{RH}(\rho) = \int_Q |\nabla \sqrt{\rho}|^2 - \int_Q G\rho + \iint_{Q \times Q} \rho(x)\, G(x - y)\, \rho(y)\, dx dy,
$$

which is still a model from the Density Functional Theory. One can prove that our guess was wrong, and that, actually, the limit of the energy per unit volume in the Hartree model is strictly less than I_{per}^{RH}.

In the Hartree model, we are led to the following conjecture. We make first the natural assumption that the electronic density becomes periodic, by passing to the thermodynamic limit (we are not able to prove this claim), and we also postulate that, roughly speaking, each electron is represented by the same electronic wavefunction φ up to a translation (which is known as a *Wannier function* in Solid State Physics literature). Within these assumptions, the periodic electronic density is given by $\rho = \sum_{k \in \mathbf{Z}^3} \varphi^2(\cdot + k)$. Our last **postulate** is then that the energy per unit volume converges to a periodic model I_{per}^H, which is given by

$$I_{per}^H = \inf \left\{ E_{per}^H(\varphi) \; ; \varphi \in H^1(\mathbf{R}^3), \int_{\mathbf{R}^3} \varphi^2 = 1 \right\} + \frac{M}{2}, \tag{9}$$

$$
\begin{aligned}
E_{per}^H(\varphi) &= \int_{\mathbf{R}^3} |\nabla \varphi|^2 - \frac{1}{2} D(\varphi^2, \varphi^2) - \int_{\mathbf{R}^3} G\varphi^2 \\
&\quad + \frac{1}{2} \iint_{\mathbf{R}^3 \times \mathbf{R}^3} \varphi^2(x) G(x-y) \varphi^2(y) \, dx dy \tag{10} \\
&= \int_{\mathbf{R}^3} |\nabla \varphi|^2 - \frac{1}{2} D(\varphi^2, \varphi^2) \\
&\quad - \int_Q G\rho + \frac{1}{2} \iint_{Q \times Q} \rho(x) G(x-y) \rho(y) \, dx dy, \tag{11}
\end{aligned}
$$

with $\rho(x) = \sum_{k \in \mathbf{Z}^3} \varphi^2(x+k)$. Let us observe at this stage that we recognize in (11) the same periodic electrostatic energy as in the TFW model.

Let us emphasize once more that we are not able to **prove** the convergence of the energy per unit volume to I_{per}^H, partly owing to the fact that the different strategies which are described in [16] only work for convex functionals. (Nevertheless, we provide in [18] an example of some Hartree model where the nuclei are smeared in a very specific manner, and for which we may establish the convergence of the energy per unit volume to the periodic model (9)–(10).) In [18], we give a proof of the well-posedness of the Hartree periodic model, in the sense that, **every** minimizing sequence of I_{per}^H admits a subsequence which is compact in $H^1(\mathbf{R}^3)$ (up to a translation along vectors which leave the crystalline lattice invariant). In particular, I_{per}^H admits a minimum φ_{per}, which is precisely the so-called Wannier function. The proof of this claim, which is given in [18], is very tricky and raises difficult mathematical questions. It is worth mentioning that the Euler-Lagrange equations satisfied by any minimizer φ_{per} of the periodic Hartree model already appear in Solid State Physics textbooks, which, in turn, gives us some confidence in the validity of our model. For the sake of consistency, let us write down here this "periodic Hartree equation" as

$$-\Delta \varphi_{per} - \left(|\varphi_{per}|^2 \star \frac{1}{|x|}\right)\varphi_{per} - G\varphi_{per} + (|\varphi_{per}|^2 \star G)\varphi + \theta \varphi_{per} = 0,$$

where $\theta > 0$ is the Lagrange multiplier.

We now turn to the next section which is devoted to Hartree-Fock type models.

4   Hartree-Fock type models

Before writing down the Hartree-Fock model, let us already say that, in all that follows, the spin effects are taken into account only through the Pauli exclusion principle; in particular, we shall only consider complex-valued wave-functions. In the absence of magnetic fields, re-incorporating the spin dependence does not affect neither our mathematical analysis nor our results.

Let us first recall the Hartree-Fock model. In this framework, the ground-state energy is defined by the following minimization problem :

$$I_{\Lambda_N}^{HF} = \inf \left\{ E_{\Lambda_N}^{HF}(\varphi_1, \ldots, \varphi_N) \; ; \varphi_i \in H^1(\mathbf{R}^3), \int_{\mathbf{R}^3} \varphi_i \, \varphi_j^* = \delta_{i,j} \right\},$$

where

$$E_{\Lambda_N}^{HF}(\varphi_1, \ldots, \varphi_N) = \sum_{i=1}^{N} \int_{\mathbf{R}^3} |\nabla \varphi_i(x)|^2 \, dx + E_{\Lambda_N}^{elec}(\rho_N)$$

$$- \frac{1}{2} \iint_{\mathbf{R}^3 \times \mathbf{R}^3} \frac{|\rho_N(x,y)|^2}{|x-y|} \, dx dy,$$

with $\rho_N(x,y) = \sum_{i=1}^{N} \varphi_i(x) \, \varphi_i^*(y)$ and $\rho_N(x) = \rho_N(x,x) = \sum_{i=1}^{N} |\varphi_i(x)|^2$.

The existence of a minimizer for the HF minimization problem is a standard fact due to Lieb and Simon [42], or Lions [43] (where a proof of the existence of infinitely many solutions to the Hartree-Fock equations is also provided).

In fact, it happens that, for the mathematical analysis of the thermodynamic limit problem of this model, it is more convenient to use an equivalent formulation of the ground-state energy in terms of density matrices. The proof that the two formulations are indeed equivalent is due to Lieb [38] (see also Bach [3] and Lions [44]).

The (reduced) one-particle density matrix, denoted by γ in the sequel, is a self-adjoint operator on $L^2(\mathbf{R}^3)$, with finite trace, such that

$$0 \leq \gamma \leq \mathbf{1}$$

(in the sense of self-adjoint operators) where $\mathbf{1}$ denotes the identity on $L^2(\mathbf{R}^3)$, and whose trace fits the number of electrons. To every such operator is associated a density ρ_γ, which is a non-negative function in $L^1(\mathbf{R}^3)$ such that

$$\mathrm{Tr} \; \gamma = \int_{\mathbf{R}^3} \rho_\gamma.$$

Indeed, the density matrix being a Hilbert-Schmidt operator on $L^2(\mathbf{R}^3)$, its kernel (still denoted by γ) may be decomposed along a complete set of orthonormal eigenfunctions $(\varphi_n)_{n\geq 1} \in L^2(\mathbf{R}^3)$ of γ associated to the eigenvalues $0 \leq \lambda_n \leq 1$, in such a way that

$$\gamma(x,y) = \sum_{n \geq 1} \lambda_n \, \varphi_n(x) \, \varphi_n^*(y).$$

The density ρ_γ is then given by

$$\rho_\gamma(x) = \sum_{n \geq 1} \lambda_n \, |\varphi_n(x)|^2.$$

The Hartree-Fock functional may then be expressed in terms of density matrices as

$$E_{\Lambda_N}^{HF}(\gamma) = \text{Tr}[-\Delta\,\gamma] + E_{\Lambda_N}^{elec}(\rho_\gamma) - \frac{1}{2}\iint_{\mathbf{R}^3\times\mathbf{R}^3}\frac{|\gamma(x,y)|^2}{|x-y|}\,dxdy,$$

where $E_{\Lambda_N}^{elec}(\rho_\gamma)$ is defined in (7). Let us note that, in the above expression of the energy, we have

$$\text{Tr}\,[-\Delta\,\gamma] = \sum_{n\geq 1}\lambda_n\int_{\mathbf{R}^3}|\nabla\varphi_n(x)|^2\,dx,$$

and that the last term, namely, $-\dfrac{1}{2}\iint_{\mathbf{R}^3\times\mathbf{R}^3}\dfrac{|\gamma(x,y)|^2}{|x-y|}\,dxdy$ is called the *exchange term*. The Hartree-Fock ground-state energy is then defined by

$$I_{\Lambda_N}^{HF} = \inf\{E_{\Lambda_N}^{HF}(\gamma)\,;\gamma\in\Gamma_N\},$$

where the set of admissible density matrices Γ_N is composed of density matrices as above, satisfying in addition that $\text{Tr}\,(\gamma) = N$ and that $\text{Tr}\,[-\Delta\,\gamma] < +\infty$ (these conditions ensure that the energy functional takes finite values on the set of constraints).

Let us now turn to the thermodynamic limit of this model, which is studied in [19] (our results have been announced in [17]). The first result is that the energy per unit volume has the expected behaviour in the limit of large volumes. As a consequence of this fact, it is checked in [19] that the exchange term also behaves linearly with respect to N as N goes to infinity. This asymptotic behaviour however admitted in the Quantum Chemistry literature (see for example [47]) was not known so far, to the best of our knowledge, except in the simplified framework of the free electron gas by Friesecke [30]. Let us emphasize the fact that the exchange term features an asymptotic behaviour which is different from the other electrostatic terms. (Recall that each of them is of the order of $N^{5/3}$ as N goes to infinity.) The advantage of the formulation in terms of density matrices is that the set Γ_N of admissible density matrices is convex. Unfortunately, the Hartree-Fock functional itself is not convex with respect to the density matrix because of the exchange term. Therefore, the strategies of proofs proposed in [16] do not apply here, and so far we have not been able to identify the thermodynamic limit of the Hartree-Fock energy per unit volume in a rigorous mathematical manner. We shall only propose in the following a periodic model which is likely to be the Hartree-Fock model for crystals. Nevertheless, before stating this model, we shall explain what kind of progress is at our reach for a simplified model, which is referred to as the Reduced Hartree-Fock model in the mathematical literature since it was introduced by Solovej in [56, 57], while it has nothing to do with the Reduced Hartree-Fock model in the sense of chemists ... This model is obtained from the Hartree-Fock model expressed in terms of density matrices, by getting rid of the exchange term. Since we have already observed that this term has a specific behaviour, by letting it apart at a first stage, we aim at understanding first the behaviour of the kinetic energy term $\text{Tr}\,[-\Delta\,\gamma]$ (since we still hope that the sum of the electrostatic terms should once more provide a known term according to (8)). Let us thus write down the RHF model in the following way

$$I_{\Lambda_N}^{RHF} = \inf\{E_{\Lambda_N}^{RHF}(\gamma); \gamma \in \Gamma_N\},$$

$$E_{\Lambda_N}^{RHF}(\gamma) = \mathrm{Tr}\,[-\Delta\,\gamma] + E_{\Lambda_N}^{elec}(\rho_\gamma),$$

and make the following crucial observation: The RHF functional is convex with respect to the density matrix, and even strictly convex with respect to the associated density ρ_γ. Moreover, as shown by Solovej in [57], for every N and Λ_N, there exists a minimizer γ_{Λ_N} of $I_{\Lambda_N}^{RHF}$, and correspondingly, a unique electronic density ρ_{Λ_N}.

Thanks to this convexity property, and by adapting the arguments which are developed in [16], we prove in [19] the thermodynamic limit of the energy per unit volume for the RHF model. In order to state the RHF periodic model, we need to extend the definition of the reduced one-particle density matrices to this periodic framework. A *periodic density matrix* γ_{per} will be defined as a self-adjoint operator on $L^2(\mathbf{R}^3)$, still satisfying $0 \le \gamma_{per} \le \mathbf{1}$, and which commutes with the group of translations $\{\tau_k\}_{k\in\mathbf{Z}^3}$ which leave invariant the underlying crystalline lattice—$\mathbf{Z}^3$ here; this property may be simply written as

$$[\tau_k, \gamma_{per}] = 0, \text{ for every } k \in \mathbf{Z}^3. \tag{12}$$

This is precisely where the meeting point between our analysis and the standard theory of Bloch waves in Solid State physics stands (see, among other references, [1, 26] and also [33, 51] for a mathematical presentation). Indeed, thanks to the commutation relations (12) satisfied by the periodic density matrix, we may extend to our operator the famous Bloch theorem in Solid State Physics, which determines the spectral decomposition of Schrödinger operators with periodic potentials in terms of Bloch waves. More precisely, Bloch's theorem states that a real number λ is in the spectrum of this kind of Schrödinger operators (*i. e.* $-\Delta + V_{per}$, with V_{per} being Q-periodic) if and only if it corresponds to a generalized eigenfunction, which is of the form of a plane wave $e^{i\xi\cdot x}$ times a periodic function, for some ξ in $\mathbf{R}^3$. (We recognize the well-known definition of *Bloch waves*.) A full description of the eigenfunctions is obtained by restricting the phase ξ to $[-\pi; +\pi[^3$, that we shall denote by $Q^\star$ in the following. (Actually, for general crystalline lattices, the phase is restricted to the dual (or reciprocal) lattice of the original lattice—see for example [10].) In other words, the spectral analysis of $-\Delta + V_{per}$ on $L^2(\mathbf{R}^3)$ reduces to the spectral analysis of the family of operators $-\Delta + V_{per}$ on $L^2_\xi(Q)$ ($\xi \in Q^\star$), where $L^2_\xi(Q)$ is the space of locally square integrable Bloch waves associated to a given phase ξ in $Q^\star$. By complete analogy with the model case of the Schrödinger operator with periodic potential, any periodic density matrix γ_{per} which enjoys the properties defined above may be diagonalized on $L^2_\xi(Q)$, one phase ξ at a time; to a fixed $\xi \in Q^\star$ corresponds a self-adjoint operator γ_ξ on $L^2_\xi(Q)$ such that

$$0 \le \gamma_\xi \le \mathbf{1}, \tag{13}$$

(where this time $\mathbf{1}$ has to be understood as the identity on $L^2_\xi(Q)$). The aim of this diagonalization of our operator γ_{per} on stable subspaces is of course to obtain a family of operators γ_ξ which are easier to diagonalize than the original one. (Let us make the obvious remark that an operator which commutes with infinitely many translations cannot be compact.) Therefore, we shall make a further assumption on the admissible periodic density matrices, which is simply that, for every ξ in $Q^\star$,

the operator γ_ξ has a finite trace on $L^2_\xi(Q)$.

Therefore, for every ξ in $Q^\star$, the Hilbert-Schmidt kernel of γ_ξ (which is still denoted by γ_ξ) may be decomposed along a complete sequence of eigenfunctions of γ_ξ, corresponding to eigenvalues $\lambda_n(\xi)$ between 0 and 1 (thanks to (13)), in the following way

$$\gamma_\xi(x,y) = \sum_{n\geq 1} \lambda_n(\xi)\, e^{-i\xi\cdot(x-y)}\, u_n(\xi;x)\, u_n^*(\xi;y)$$

where $u_n(\xi;x)$ is a hilbertian basis of $L^2(Q)$ consisting of Q-periodic functions. At each fixed phase, we may define a density $\rho_\xi(x)$ by

$$\rho_\xi(x) = \sum_{n\geq 1} \lambda_n(\xi)\, |u_n(\xi;x)|^2.$$

This function is clearly Q-periodic and satisfies $\int_Q \rho_\xi(x)\, dx = \sum_{n\geq 1} \lambda_n(\xi) = \mathrm{Tr}_{L^2_\xi(Q)}\gamma_\xi$. The full periodic density matrix γ_{per} is then recovered from the family of γ_ξ's through the relation

$$\gamma_{per}(x,y) = \frac{1}{(2\pi)^3} \int_{Q^\star} \gamma_\xi(x,y)\, d\xi,$$

and the periodic electronic density is given by

$$\rho_{per}(x) = \frac{1}{(2\pi)^3} \int_{Q^\star} \rho(\xi,x)\, d\xi.$$

(Let us also mention [47] where one can find definitions of periodic density matrices which are similar to ours.) The condition that each cell contains exactly one electron is now written in different equivalent manners as

$$1 = \int_Q \rho_{per} = \frac{1}{(2\pi)^3} \int_{Q^\star} \sum_{n\geq 1} \lambda_n(\xi)\, d\xi = \frac{1}{(2\pi)^3} \iint_{Q\times Q^\star} \rho_\xi(x)\, d\xi dx.$$

An extra condition on the γ_ξ's is necessary in order to give a sense to the analogue of the kinetic energy term in the periodic setting. This condition is simply $\int_{Q^\star} \mathrm{Tr}_{L^2_\xi(Q)}[-\Delta_\xi \gamma_\xi]\, d\xi < +\infty$, where the notation $-\Delta_\xi$ stands for the Laplacian with quasi-periodic boundary conditions on the boundary of the cell. Finally, we denote by Γ, the set of periodic one-particle density matrices which fulfill the series of properties required before.

With these "new" objects at hand, we are able to define the RHF periodic model as

$$I_{per}^{RHF} = \inf\{E_{per}^{RHF}(\gamma_{per}); \gamma_{per} \in \Gamma\},$$

$$E_{per}^{RHF}(\gamma_{per}) = \frac{1}{(2\pi)^3} \int_{Q^*} \mathrm{Tr}_{L_\xi^2(Q)}[-\Delta_\xi \gamma_\xi]\, d\xi$$

$$- \int_Q G\, \rho_{per} + \frac{1}{2} \iint_{Q\times Q} \rho_{per}(x)\, G(x-y)\, \rho_{per}(y)\, dx\, dy.$$

We prove in [19] that this periodic minimization problem admits a minimum $\gamma \in \Gamma$, which corresponds to a unique minimizing density $\rho(x)$ (thanks to the strict convexity of the periodic energy functional with respect to the density). Moreover, we also establish there that this model is obtained by passing to the thermodynamic limit in the RHF energy per unit volume.

Let us return now to the Hartree-Fock model. In this setting, we are not able to prove rigorously that the electronic density becomes periodic, even if it should be true in the absence of magnetic fields. (When magnetic fields are present, and within models with spins, one may observe periodicity on a larger cell; see [47] and [5].) However, we **postulate** that, should it indeed be the case, the Hartree-Fock energy per unit volume would converge to the following periodic model. :

$$I_{per}^{HF} = \inf\{E_{per}^{HF}(\gamma_{per}); \gamma_{per} \in \Gamma\},$$

$$E_{per}^{HF}(\gamma_{per}) = \frac{1}{(2\pi)^3} \int_{Q^*} \mathrm{Tr}_{L_\xi^2(Q)}[-\Delta_\xi \gamma_\xi]\, d\xi - \int_Q G\, \rho_{per}$$

$$+ \frac{1}{2} \iint_{Q\times Q} \rho_{per}(x)\, G(x-y)\, \rho_{per}(y)\, dx\, dy - \frac{1}{2} \iint_{Q\times\mathbf{R}^3} \frac{|\gamma_{per}(x,y)|^2}{|x-y|}\, dx\, dy.$$

Comparing with the analogue periodic model in the RHF setting, the only new term is the exchange term $-\frac{1}{2} \iint_{Q\times\mathbf{R}^3} \frac{|\gamma_{per}(x,y)|^2}{|x-y|}\, dx\, dy$.

We prove in [19] that this Hartree-Fock periodic model is well-posed; in particular, I_{per}^{HF} admits a minimizer γ in Γ. Before writing down the HF periodic equations, we claim that the exchange term may be rewritten in the equivalent form

$$\iint_{Q\times\mathbf{R}^3} \frac{|\gamma_{per}(x,y)|^2}{|x-y|}\, dx\, dy$$

$$= \frac{1}{(2\pi)^6} \iiiint_{(Q^*)^2\times Q^2} \gamma_\xi(x,y)\, W_\infty(\xi-\xi', x-y)\, \gamma_{\xi'}(x,y)^*\, d\xi\, d\xi'\, dx\, dy,$$

where, for every η in Q^* and z in $\mathbf{R}^3$,

$$W_\infty(\eta, z) = \sum_{k\in\mathbf{Z}^3} \frac{e^{ik\cdot\eta}}{|z+k|}.$$

(Note that $e^{i\eta\cdot z} W_\infty(\eta, z)$ is Q-periodic with respect to z.) On this latter form the non-local feature of the exchange term is more easily seen. The Euler-Lagrange equation satisfied by any minimizer may be written as follows, at every fixed phase ξ in Q^*, and for every $n \geq 1$,

$$-(i\nabla + \xi)^2 u_n(\xi,\cdot) - G\, u_n(\xi,\cdot) + \left(\int_Q G(\cdot - y)\, \rho_\xi(y)\, dy\right) u_n(\xi,\cdot)$$

$$- \frac{1}{(2\pi)^3} \iint_{Q^*\times Q} \gamma(\xi';\cdot;y)\, W_\infty(\xi-\xi';\cdot - y)\, e^{i\xi\cdot y} u_n(\xi, y)\, dy\, d\xi'$$

$$= \varepsilon_n(\xi)\, u_n(\xi,\cdot), \quad \text{on } Q,$$

together with

$$\begin{cases} \lambda_n(\xi) = 0 \implies \varepsilon_n(\xi) \geq \pi, \\ 0 < \lambda_n(\xi) < 1 \implies \varepsilon_n(\xi) = \pi, \\ \lambda_n(\xi) = 1 \implies \varepsilon_n(\xi) \leq \pi, \end{cases}$$

where the real number π enters the equations as a Lagrange multiplier. It is worth noticing that these equations seem to be already known in Solid State Physics literature, thereby strengthening our feeling that our model should be the correct one.

We still do not know whether any minimizer of our periodic model is a projector on the lowest eigenvalues of the Hartree-Fock periodic operator for each phase, as it is assumed to be the case for insulators (see the paper by Resta in [47]), and as it is known to be the case in the molecular HF model (see [4] and [43]).

5   The optimal periodic configuration for the nuclei

Hitherto we have concentrated our efforts on the first stage in the Born-Oppenheimer approximation, which consists in fixing *a priori* the locations of the nuclei, and then in determining the electronic ground-state energy, which is therefore parameterized by the nuclear configuration. We were also considering a finite number N of nuclei which were fixed at the points of a given periodic lattice, and we were interested in the asymptotic behaviour of quantities which describe the electronic state, like the electronic densities, and the ground-state energy per unit volume, when N goes to infinity and the nuclei fill in the entire periodic lattice. A question of great importance (but of outstanding mathematical difficulty) is the following. The second fundamental stage within the Born-Oppenheimer approximation consists next in minimizing the ground-state energy of the electrons with respect to all possible geometries of nuclei; the nuclei are therefore no more constrained to lie at points of a lattice, and we then obtain an "absolute ground-state", taking into account the less-energy configuration for both the electrons and the nuclei. It is of course tempting to pass to the thermodynamic limit on the absolute ground-state energy per unit volume, and to raise the following issues. Being given a model within which such an optimal geometry exists for the nuclei (like TFW or Hartree—see [20]), what is the behaviour of this optimal geometry when the number of nuclei goes to infinity ? Does it become periodic or do the nuclei arrange according to some ordered structures (think of quasi-crystals for example) ? This would explain why crystals are observed to have a tendency to form at absolute zero temperature. Some important preliminary results on the so-called crystal problem are available within some models for classical atoms in one or two dimensions (see, for example [31, 49, 50, 60]). In both cases, it is shown that the optimal configuration for the nuclei becomes periodic in the thermodynamic limit and, for two dimensional studies, the shape of the optimal cell is well-identified. Owing to the well-known Teller's no-binding theorem in the Thomas-Fermi setting (see [58, 6], and [41], for a rigorous mathematical proof), the simplest quantum models within which binding occurs are the Hartree and the TFW model (see [20]). Unfortunately, as far as we know, analogous results for quantum models are beyond one's reach, for the moment. However, we shall report in this section on recent works by Blanc and one of us [9, 11, 12] in the framework of the TFW model which may be seen as preliminary steps in this direction. (In the following, most of the notation is inherited from [12].) Indeed, one may define the periodic TFW model for general crystals, and not simply for a unit cubic cell with one point nucleus of unit charge, in the following way (see [16] and [12]). For every proper periodic lattice ℓ of $\mathbf{R}^3$, namely

$$\ell = \{k\,a_1 + l\,a_2 + m\,a_3; \ (k,l,m) \in \mathbf{Z}^3\},$$

where the three vectors a_i, $i = 1,2,3$, span $\mathbf{R}^3$, a primitive cell of the crystal is defined by

$$\Gamma(\ell) = \left\{ x\,a_1 + y\,a_2 + z\,a_3; \ (x,y,z) \in [-\frac{1}{2};\frac{1}{2}[\right\}.$$

We assume that one nucleus of unit charge lie at the center of each cell. The set of all proper lattices of $\mathbf{R}^3$ is denoted by $\mathcal{L}$. According to the results in [16], and by analogy with (3), (4), (5) and (6), the ground-state energy of the associated neutral crystal is given by

$$I(\ell) = \inf\left\{ E_\ell(\rho); \ \rho \geq 0, \ \sqrt{\rho} \in H^1_{per}(\Gamma(\ell)), \ \int_{\Gamma(\ell)} \rho = 1 \right\}, \tag{14}$$

where

$$E_\ell(\rho) = \int_{\Gamma(\ell)} |\nabla\sqrt{\rho}|^2 + \int_{\Gamma(\ell)} \rho^{5/3} - \int_{\Gamma(\ell)} G_\ell(x)\,\rho(x)\,dx$$
$$+ \ \frac{1}{2}\iint_{\Gamma(\ell)\times\Gamma(\ell)} \rho(x)\rho(y)G_\ell(x - y)dxdy, \tag{15}$$

with $H^1_{per}(\Gamma(\ell))$ being the subset of $H^1_{loc}(\mathbf{R}^3)$ consisting of functions which satisfy the periodic boundary conditions on the boundary of $\Gamma(\ell)$. The potential G_ℓ which appears in the definition (15) of the TFW functional is defined, in a unique way, by

$$-\Delta G_\ell = 4\pi \left[\sum_{k\in\ell} \delta(\cdot - k) - \frac{1}{|\Gamma(\ell)|} \right], \tag{16}$$

and

$$\lim_{|x|\to 0} \left[G_\ell(x) - \frac{1}{|x|} \right] = 0, \tag{17}$$

where $|\Gamma(\ell)|$ is simply the volume of $\Gamma(\ell)$. (Note that G_ℓ is actually the Green function of the periodic Laplacian on $\Gamma(\ell)$, and that the normalization (17) on G_ℓ which is made in [9, 12], instead of $\int_{\Gamma(\ell)} G_\ell = 0$, simply helps for the calculations.) The problem under consideration in [9] consists in looking at the periodic ground-state energy (14)-(15) as a function of the primitive cell ℓ, and next in searching for a primitive cell which yields the smallest possible periodic ground-state energy in the framework of the TFW model. In other words, it is shown that every minimizing sequence $(\ell_n)_{n\geq 1}$ of the following minimization problem

$$I = \inf\{I(\ell); \ \ell \in \mathcal{L}\}$$

admits a subsequence which converges to some optimal cell. Interesting questions still remain to be solved on the uniqueness and the geometric properties (symmetries for instance) of this cell. The strategy of proof deserves some comments. Assuming that $(\ell_n)_{n\geq 1}$ is a minimizing sequence for I (*i.e.* $(\ell_n)_{n\geq 1}$ is a sequence of proper lattices of $\mathbf{R}^3$, and $\lim_{n\to+\infty} I(\ell_n) = I$), the point consists in verifying that $\ell_n = (a_1^n; a_2^n; a_3^n)$ converges to some proper lattice of $\mathbf{R}^3$. Any proper lattice ℓ_n is completely determined by the following data. We shall denote by R_i^n the respective length of a_i^n, for $i = 1$, 2 or 3, and by $\vartheta_{i,j}^n$ the angle between a_i^n and a_j^n, for $i \neq j$, and we shall assume, without loss of generality that

$$R_1^n \leq R_2^n \leq R_3^n. \tag{18}$$

The first remark is borrowed by Blanc from crystallography textbooks, where one observes that the angles of proper lattices may be chosen in the compact segment $\left[\frac{\pi}{3}; \frac{\pi}{2}\right]$. Therefore, extracting a subsequence, if necessary, we already dispose of the convergence of the angles as n goes to infinity. It remains to check that the so-called radii R_i^n also converge (at least up to a further subsequence) to some positive quantities R_i, for $i = 1$, 2 or 3. The first condition which has to be satisfied by the minimizing sequence is that R_1^n stays away from zero as n goes to infinity (and thus so do R_2^n and R_3^n, thanks to (18)). Otherwise, the minimizing cells shrink to a single point as n goes to infinity, and there is no cell remaining at all after passing to the limit. This condition is ensured by Teller's theorem for the TF energy which is an obvious bound from below for the TFW energy (see [9]). Secondly, a series of "stability type inequalities" have to be checked, which have a simple physical interpretation : They translate into mathematical language the fact that one has to bring energy to the system under consideration in order to increase the size of the cell to infinity (that is, some of the radii R_i^n to infinity). Owing to (18), such a situation may arise in three different manners. First of all, in order to avoid the case when R_1^n goes to infinity as n goes to infinity (and thus so do R_2^n and R_3^n), Blanc shows that the absolute periodic ground-state I corresponds to an energy which is strictly below the ground-state energy of a single isolated (neutral) atom. Secondly, in order to exclude the two other cases, namely the case when R_2^n (and thus R_3^n) goes to infinity, on the one hand, and the case when R_3^n goes to infinity on the other hand, Blanc checks that the absolute periodic ground-state I corresponds to an energy which is strictly below the ground-state energy of lineic polymers, on the one hand, and of thin films on the other hand. The TFW models for the ground-state energy of the lineic polymers, denoted by I_{pol}^{TFW}, and of the thin films, denoted by I_{film}^{TFW}, have been introduced by Blanc and one of us in [11]. They have been obtained from a thermodynamic limit strategy by extending the methods of [16] to the case of "infinite molecules", where the nuclei are arranged periodically along a line in a case of a lineic polymer or on a plane in the case of thin films. Let us now write down these models. Let ℓ be a one-dimensional periodic lattice of $\mathbf{R}^3$ which is defined by the vector e in $\mathbf{R}^3$. Then the ground-state energy of the lineic polymer whose nuclei stand at the points of $\mathbf{Z}\,e$ is given by

$$I_{pol}^{TFW}(\ell) = \inf\left\{E_\ell(\rho);\ \rho \geq 0,\ \int_{\Gamma(\ell)} \rho = 1,\right.$$

$$\sqrt{\rho} \in H_{per}^1(\Gamma(\ell)), \ \int_{\Gamma(\ell)} \log\left(2 + |x|\right) \rho < +\infty \Bigg\},$$

with

$$\Gamma(\ell) = \mathbf{R}^2 \times \left\{ t\,e;\ t \in [-\frac{1}{2}; -\frac{1}{2}[\right\},$$

and where the expression of $E_\ell(\rho)$ is the same as in (15), except for the fact that the potential G_ℓ is this time defined as a solution to

$$-\Delta G_\ell = 4\pi \sum_{k \in \mathbf{Z}} \delta(\cdot - k\,e)$$

which is periodic in the direction of e (and with additional normalization conditions which are made precise in [11]). The model for thin films is similarly written as :

$$I_{film}^{TFW}(\ell)$$

$$= \ \inf\left\{ E_\ell(\rho);\ \rho \geq 0,\ \sqrt{\rho} \in H_{per}^1(\Gamma(\ell)),\ \int_{\Gamma(\ell)} |x|\,\rho < +\infty,\ \int_{\Gamma(\ell)} \rho = 1 \right\},$$

where ℓ is a periodic lattice in $\mathbf{R}^2$, defined by two linearly independent vectors e_1 and e_2 and

$$\Gamma(\ell) = \mathbf{R} \times \left\{ t\,e_1 + s\,e_2;\ t,\ s \in [-\frac{1}{2}; -\frac{1}{2}[\right\}.$$

The corresponding potential G_ℓ which appears in the general definition (15) of the energy functional is a solution to

$$-\Delta G_\ell = 4\pi \sum_{(k;\,l) \in \mathbf{Z}^2} \delta(\cdot - k\,e_1 - l\,e_2)$$

which shares the periodicity of the two-dimensional lattice (further normalization conditions are also made precise in [11]).

In order to stimulate future work in this topic, we shall conclude this review by indicating in the forthcoming section some natural extensions of our work.

6   Extensions and perspectives

Let us recall from the introduction that, by applying the thermodynamic limit strategy to various models from Quantum Chemistry, we aimed at building new quantum-mechanical models for crystals. We have illustrated the efficiency of this method on a wide variety of models, from the toy model which is a Thomas-Fermi type model to the more sophisticated Hartree-Fock model. In any cases, this strategy provides a rigorous derivation of expressions for the ground-state energies. Moreover, since, for standard models like Hartree and Hartree-Fock models, the corresponding Euler-Lagrange equations coincide with well-known equations in Solid State Physics, we are founded to think of applying the same strategy in order to construct new models. For example, by taking our Hartree-Fock periodic model as a starting point, it should be possible (while still much more difficult) to propose a Dirac-Fock type model for the crystalline phase, in order to take into account the relativistic effects (see the paper by Dolbeault, Esteban and Séré [23] for a presentation on the Dirac-Fock model).

Besides, we may now try to extend our models by allowing for more general geometries for the nuclei. In the framework of the Thomas-Fermi-von Weizsäcker model, we have already treated in [16] the case of general crystals, which may possibly contain local defects, like impurities, or, even some cases of non-periodic structures. We give in [16] some mathematical background which should allow to define a ground-state energy for quasi-crystals in a Thomas-Fermi type setting at a first stage, and next in the Hartree or even in the Hartree-Fock setting, within different mathematical definitions of quasi-crystals (see [2, 53]). In the same spirit of taking into account more general configurations for the nuclei, and still for the TFW model, we recall from the previous section that Blanc and one of us made use of the thermodynamic limit strategy to build models for polymers and two-dimensional films. Therefore, the general method exposed in [16] seems to allow the construction of ground-state energies for arrangements of infinitely many nuclei, which may be much more complicated than the model case of the crystalline lattice.

As also mentioned in the previous section, it is very challenging to tackle the thermodynamic limit problem of quantum models when the nuclei no more stand at the point of a periodic lattice but are rather disposed accordingly to a configuration where they minimize the ground-state energy. By understanding the behaviour of this optimal geometry when the number of nuclei goes to infinity, one tries to contribute to a theoretical explanation to the empirical observation that crystals forms at absolute zero temperature.

Eventually, having different models for the solid phase at our disposal, it would also be interesting to make use of them to build models for the macroscopic scale, with the help of homogenization techniques, and to calculate macroscopic quantities, say, for instance, elasticity constants. Besides their intrinsic interest, by addressing these natural questions, we also have in mind to validate a bit further our models.

Of course, beyond the mathematical analysis of the various equations we may obtain, we also intend to give a sound ground to numerical computations based on them. We shall not develop further on the important question of the numerical aspects of the modeling of the solid phase, and we would rather refer the reader to Blanc [10].

Bibliography

[1] N. W. Ashcroft & N. D. Mermin, *Solid-state Physics*, Saunders College Publishing, 1976.

[2] F. Axel & D. Gratias ed., *Beyond Quasicrystals*, Centre de Physique Les Houches, Les Editions de Physique, Springer, 1995.

[3] V. Bach, *Error bound for the Hartree-Fock energy of atoms and molecules*, Comm. Math. Phys.,147, 1992, pp. 527-548.

[4] V. Bach, E.H. Lieb, M. Loss & J.P. Solovej, *There are no unfilled shells in Hartree-Fock theory*, Phys. Rev. Lett., 72 (19), 1994, pp. 2981-2983.

[5] V. Bach, E. H. Lieb & J. P. Solovej, *Generalized Hartree-Fock Theory and the Hubbard Model*, J. Stat. Phys. 76, 1994, pp. 3-90.

[6] N. Balàzs, *Formation of stable molecules within the statistical theory of atoms*, Phys. Rev., 156, pp. 42-47, 1967.

[7] R. Balian, *From microphysics to macrophysics ; methods and applications of Statistical Physics*, I & II, Springer Verlag, 1991.

[8] R. Benguria, H. Brézis & E. H. Lieb, *The Thomas–Fermi–von Weizsäcker theory of atoms and molecules*, Comm. Math. Phys., 79, 1981, pp. 167-180.

[9] X. Blanc, *Geometry optimization for crystals in Thomas–Fermi type theories of solids*, submitted to Comm. Partial Differential Equations. Also available at : http://cermics.enpc.fr/reports/CERMICS-99-173.ps.gz.

[10] X. Blanc, *A mathematical insight ab initio simulations of the solid phase*, in this volume.

[11] X. Blanc & C. Le Bris, *Optimisation de géométrie dans le cadre des théories de Thomas-Fermi pour les cristaux périodiques [Geometry optimization for Thomas-Fermi type theories of solids]*, C. R. Acad. Sci. Paris Sér. I Math., t. 329, 1999, pp. 551-556.

[12] X. Blanc & C. Le Bris, *Thomas-Fermi type models for polymers and thin films*, to appear in Advances in Differential Equations. Also available at : http://cermics.enpc.fr/reports/CERMICS-99-164.ps.gz

[13] J. Callaway, *Quantum Theory of the Solid State*, Academic Press, 1974.

[14] E. Cancès, M. Defranceschi & C. Le Bris, *Some recent mathematical contributions to Quantum Chemistry*, Int. J. Quantum Chem., 74, 1999, pp. 553-557.

[15] I. Catto, C. Le Bris & P.-L. Lions, *Limite thermodynamique pour des modèles de type Thomas-Fermi [Thermodynamic limit for Thomas-Fermi type models]*, C. R. Acad. Sci. Paris Sér. I Math., t. 322, pp. 357-364, 1996.

[16] I. Catto, C. Le Bris & P.-L. Lions, *Mathematical Theory of thermodynamic limits : Thomas-Fermi type models*, Oxford University Press, 1998.

[17] I. Catto, C. Le Bris & P.-L. Lions, *Sur la limite thermodynamique pour des modèles de type Hartree et Hartree-Fock [On the thermodynamic limit for Hartree and Hartree-Fock type models]*, C. R. Acad. Sci. Paris Sér. I Math., t. 327, 1998, pp. 259-266.

[18] I. Catto, C. Le Bris & P.-L. Lions, *On some periodic Hartree-type models for crystals*, available at : http://www.math.utexas.edu/mp_arc/c/99/99-392.ps.gz

[19] I. Catto, C. Le Bris & P.-L. Lions, *On the thermodynamic limit for Hartree-Fock type models*, available at : http://www.math.utexas.edu/mp_arc/c/00/00-93.ps.gz

[20] I. Catto & P.-L. Lions, *Binding of atoms and stability of molecules in Hartree and Thomas-Fermi type theories*, Parts I-IV, Comm. Partial Differential Equations, 17 &18, 1992 & 1993.

[21] M. Defranceschi & C. Le Bris, *Computing a molecule : A mathematical viewpoint*, J. Math. Chem., 21, No.1, 1997, pp. 1-30.

[22] M. Defranceschi & C. Le Bris, *Computing a molecule in its environment : A mathematical viewpoint*, Int. J. Quantum Chem., 71, 1999, pp. 227-250.

[23] J. Dolbeault, M. Esteban & E. Séré, *Variational methods in relativistic quantum mechanics : new approach to the computation of Dirac eigenvalues*, in this volume.

[24] R. Dovesi, C. Freyria-Fava, C. Roetti C., L. Salasco & V.R. Saunders, *On the electrostatic potential in crystalline systems where the charge density is expanded in Gaussian functions*, Mol. Phys., 77, 1992, pp. 629-665.

[25] R.M. Dreizler & E.K.U. Gross, *Density functional theory*, Springer-Verlag, 1990.

[26] M.S.P. Eastham, *The Spectral Theory of Periodic Differential Equations*, Scottish Acad. Press, Edinburgh-London, 1973.

[27] P. P. Ewald, Ann. Phys. (Leipzig) 64, 1921, pp. 253.

[28] C. Fefferman, *The thermodynamic limit for a crystal*, Comm. Math. Phys., 98, 1985, pp. 289-311.

[29] C. Fefferman, *The atomic and molecular nature of matter*, Rev. Mat. Iberoamericana, vol. 1, 1, 1985.

[30] G. Friesecke, *Pair Correlations and Exchange Phenomena in the Free Electron Gas*, Comm. Math. Phys. 184, 1997, pp. 143-171.

[31] C. S. Gardner & Ch. Radin, *The infinite-volume ground state of the Lennard-Jones potential*, J. Stat. Phys., 20, No. 6, 1979, pp. 719-724.

[32] D. Hartree, *The wave-mechanics of an atom with a non-Coulomb central field. Part I. Theory and methods*, Proc. Camb. Phil. Soc. 24, 1928, pp. 89-132.

[33] Y. E. Karpeshina, *Perturbation theory for the Schrödinger operator with a periodic potential*, Lecture Notes in Mathematics 1663, Springer-Verlag, 1997.

[34] C. Kittel, *Introduction to Solid-State Physics*, 6th Edition, Wiley and Sons, 1986.

[35] J. L. Lebowitz & E. H. Lieb, *Existence of thermodynamics for real matter with Coulomb forces*, Phys. Rev. Lett., 22, 13, 1969, pp. 631-634.

[36] E. H. Lieb, *The stability of matter : from atoms to stars*, Bull. Amer. Math. Soc., 22, 1, 1990, pp. 1-49.

[37] E. H. Lieb, *Thomas-Fermi and related theories of atoms and molecules*, Rev. Modern Phys., 53, 4, 1981, pp. 603-641.

[38] E. H. Lieb, *A Variational principle for many-fermion systems*, Phys. Rev. Lett. 46, 1981, pp. 457-459. *Errata* ibid 47, 1981, pp. 69.

[39] E. H. Lieb & J.L. Lebowitz, *The constitution of matter : existence of thermodynamics for systems composed of electrons and nuclei*, Adv. Math., 9, 1972, pp. 316-398.

[40] E. H. Lieb & J.L. Lebowitz, *Lectures on the thermodynamic limit for Coulomb systems*, Springer Lecture Notes in Physics, 20, 1973, pp. 136-161.

[41] E. H. Lieb & B. Simon, *The Thomas-Fermi theory of atoms, molecules and solids*, Adv. Math., 23, 1977, pp. 22-116.

[42] E. H. Lieb & B. Simon, *The Hartree-Fock theory for Coulomb systems*, Comm. Math. Phys., 53, 1977, pp. 185-194.

[43] P.-L. Lions, *Solutions of Hartree-Fock equations for Coulomb systems*, Comm. Math. Phys., 109, 1987, pp. 33-97.

[44] P.-L. Lions, *Hartree-Fock and related equations*, in : Nonlinear Partial Differential Equations and their Applications, Lect. Collège de France Seminar, Vol. IX, Paris, 1985-86, Pitman Res. Notes Math. Ser. 181, 1988, pp. 304-333.

[45] O. Madelung, *Introduction to Solid-state theory*, Solid State Sciences 2, Springer-Verlag, Berlin, 1981.

[46] R.G. Parr & W. Yang, *Density-functional theory of atoms and molecules*, Oxford University Press, Oxford & New York, 1989.

[47] C. Pisani ed., *Quantum Mechanical Ab Initio Calculation of the properties of crystalline materials*, Lecture Notes in Chemistry 67, Springer, 1996.

[48] Ch. M. Quinn, *An introduction to the Quantum Theory of Solids*, Clarendon Press, Oxford, 1973.

[49] C. Radin, *The ground state for soft disks*, J. Stat. Phys., 26, No. 2, 1981, pp. 365-373.

[50] C. Radin & L. S. Schulman, *Periodicity of Classical Ground-States*, Phys. Rev. Lett., 51, No. 8, 1983, pp. 621-622.

[51] M. Reed & B. Simon, *Methods of Modern Mathematical Physics*, IV: *Analysis of Operators*, Academic Press, New-York, 1978.

[52] D. Ruelle, *Statistical Mechanics : Rigorous Results*, Benjamin, New-York, 1969 and Advanced Books Classics, Addison-Wesley, 1989.

[53] M. Senechal, *Quasicrystals and Geometry*, Cambridge University Press, 1995.

[54] J. C. Slater, *Quantum Theory of Molecules and Solids*, Mac Graw Hill, 1963.

[55] J. C. Slater, *Symmetry and Energy Bands in Crystals*, Dover, 1972.

[56] J. P. Solovej, *An improvement on stability of matter in mean field theory*, In. Differential Equations and Mathematical Physics, Proceedings of the International Conference, Univ. of Alabama, Birmingham, March 1994, International Press 1995.

[57] J. P. Solovej, *Proof of the ionization conjecture in a reduced Hartree-Fock model*, Invent. Math. 104, 1991, pp. 291-311.

[58] E. Teller, *On the stability of molecules in the Thomas-Fermi theory*, Rev. Modern Phys., 34, 1962, pp. 627.

[59] R. C. Tolman, *The Principles of Statistical Mechanics*, Oxford University Press, 1962.

[60] W. J. Ventevogel, *On the configuration of a one-dimensional system of interacting particles with minimum potential energy per particle*, Physica 92 A, 1978, pp. 343-361.

[61] J. Ziman, *Principles of the Theory of Solids*, 2nd Edition, Cambridge University Press, 1972.

Chapter 6

Local density approximations for the energy of a periodic Coulomb model

O. Bokanowski [1], B. Grébert [2] and N. J. Mauser [3]

[1] *LSMA (Univ. Paris 7) and LAN (Univ. Paris 6);*
B.P. 7012, Université Paris 7,
16 rue Clisson,
F-75013 Paris, France
boka@math.jussieu.fr

[2] *Département de Mathématiques,*
Université de Nantes.
2, rue de la Houssinière,
F-44072 NANTES Cedex 03, France.
Benoit.Grebert@math.univ-nantes.fr

[3] *Inst. f. Mathematik, Univ Wien,*
Strudlhofg. 4, A–1090 Wien, Austria, and
Courant Institute,
251 Mercer Street,
NY-10012-1185 NY, USA
mauser@cma.univie.ac.at

Abstract: We deal with local density approximations for the kinetic and exchange energy terms of a periodic Coulomb model. For the kinetic energy, we give a rigorous derivation of the usual combination of the von-Weizsäcker term and the Thomas-Fermi term in the "high density" limit. Furthermore, we justify the inclusion of the Dirac term for the exchange energy and the Slater term for the local exchange potential. Our method is based on deformations (local scaling transformations) of plane waves in a periodic box.

1   Introduction

The aim of this work is to provide a simple rigorous derivation of Thomas-Fermi-von Weizsäcker and Thomas-Fermi-von Weizsäcker-Dirac models [20] in a crystal. We use the method of deformations (local scaling transformations) [12] of wave functions of constant electron density in the "high density limit" ([19], [16], [9]), i.e. the limit $N \to \infty$ at constant volume of a (periodic) box. To this end we consider a cubic crystal with N electrons and P nuclei in the elementary cell $\Omega = [0, L[^3$ (L fixed). Following [10], [11] we regard the periodic Hamiltonian:

$$H_{per} := -\sum_{i=1}^{N} \Delta_i + \sum_{i=1}^{N} V_{ext}(x_i) + \sum_{i<j} G(x_i - x_j) \tag{1}$$

where the $x_i \in \Omega$ denote the fermion positions, $V_{ext}(x) = -\sum_{l=1}^{P} Z_l G(x - R_l)$, Z_l and R_l are respectively the charges and the positions of the P kernels. Notice that the $1/|x|$ function of the Coulomb interaction has been replaced by the "periodized version" $G(x)$ of e.g. [19], which is the periodic solution of the equation $-4\pi \Delta G = \sum_{k \in L\mathbf{Z}^3} \delta(\cdot - k) - 1/L^3$, satisfying $\int_\Omega G = 0$.

As the Hamiltonian is periodic, the N-particle wavefunctions are Bloch functions [1], i.e. a shift for a lattice vector γ results in a mere phase factor : $\Psi(x+\gamma) = e^{i\Theta \cdot \gamma} \Psi(x)$ for all $x \in \mathbf{R}^{3N}$ and $\gamma \in L\mathbf{Z}^{3N}$ and for some Θ fixed in $[0, 2\pi/L[^{3N}$. On the other hand, the N-particle wavefunctions have to be antisymmetric by the Pauli principle. This implies e.g. that the Bloch phase factor has N identical components : $\Theta = (\theta, ..., \theta)$. These conditions [1] lead us to consider the following space of wave functions

$$\Lambda := \bigoplus_{\theta \in [0, 2\pi/L[^3} \Lambda_\theta \tag{2}$$

where

$$\Lambda_\theta := \{\Psi(x_1, ..., x_N) = e^{i\theta \cdot (x_1 + \cdots + x_N)} \Phi(x_1, ..., x_N), \ \Phi \in H_a^1((\mathbf{R}^3/L\mathbf{Z}^3)^N)\}.$$

(The index "a" means that the functions Φ are antisymmetric.) The energy $E(\Psi)$ and the kinetic energy $E_{kin}(\Psi)$ of a wave function Ψ are given by

$$E(\Psi) := \langle H_{per}\Psi, \ \Psi \rangle, \tag{3}$$

where $\langle , \rangle$ denotes the Hermitian scalar product in $L^2(\Omega^N)$ and

$$E_{kin}(\Psi) := \int_{\Omega^N} |\nabla \Psi(x)|^2 dx. \tag{4}$$

The fundamental energy of this N fermions system is then given by

$$E_0 := \min\{\langle H_{per}\Psi, \ \Psi \rangle, \ \Psi \in \Lambda, \ ||\Psi||_{L^2} = 1\}. \tag{5}$$

---

[1]In order to simplify the presentation we do not explicitly consider the spin.

The aim of the density functional theory is to replace the minimisation problem (5) by a minimisation problem with respect to the density ρ associated to the wave function Ψ :

$$\rho_\Psi(x) := N \int_{\Omega^{N-1}} |\Psi(x, x_2, \ldots, x_N)|^2 dx_2 \ldots dx_N. \tag{6}$$

This procedure reduces the number of variables from $3N$ to 3.

We define density functionals corresponding to the energy and the kinetic energy terms as follows,

$$
\begin{aligned}
\mathcal{E}(\rho) &:= \inf\{E(\Psi), \ \Psi \in \Lambda \text{ and } \rho_\Psi = \rho\}, \\
\mathcal{E}_{kin}(\rho) &:= \inf\{E_{kin}(\Psi), \ \Psi \in \Lambda \text{ and } \rho_\Psi = \rho\}.
\end{aligned} \tag{7}
$$

We introduce the constant $\rho_0 := N/L^3$ and the function

$$\epsilon_\rho(x) := \frac{\rho(x) - \rho_0}{\rho_0} \tag{8}$$

which describes the difference of the constant density ρ_0 (of N plane waves in a box with lenght L) and the actual density $\rho(x)$.

For densities we use the space $(0 < \alpha < 1)$

$$D_\alpha := \{\rho \in C^{0,\alpha}(\mathbf{R}^3/L\mathbf{Z}^3), \ \rho > 0, \ \int_\Omega \rho = N, \ \sqrt{\rho} \in H^1_{loc}(\mathbf{R}^3)\} \tag{9}$$

where $C^{0,\alpha}$ denotes the space of Hölderian functions of degree α endowed with the classical Hölderian norm which we will denote $\|.\|_{0,\alpha}$. By $\mathbf{R}^3/L\mathbf{Z}^3$ we denote the torus, i.e. Ω with periodic boundary conditions.

In this paper we give estimates under the "high density limit" assumption, that is in the $N \to \infty$ limit, when ϵ_ρ is small in the $C^{0,\alpha}$ norm (i.e. $\rho \sim \rho_0$). Part of the results was anounced in [6], the proofs are given in [8] (see also [7]).

In Sec 3. we give an estimate of $\mathcal{E}_{kin}(\rho)$, the kinetic energy, in Sec 4. we give an estimate on the exchange potential as it appears in the so called "$X\alpha$ method" (cf [22], which in turn yields an estimate for the Coulomb energy term (Exchange energy). This allows us finally to obtain, in Sec 5., a global energy estimate for $\mathcal{E}(\rho)$, in terms of the density function ρ. These energies are related to well known density functional energies in Quantum Chemistry and Quantum Physics, such as the Thomas-Fermi Von-Weizsäcker functional, Dirac exchange approximation, that appears in Density functional methods [22].

In the following Sec 2. we first give a lemma for the existence of a deformation as a solution of the "Jacobian problem", which is central for our estimates.

2 Deformations and the Jacobian problem

The basic idea of the deformation method is to find a local scaling transformation - the "deformation" $\mathbf{f}$, that transforms a density into another density. Hence the crucial mathematical problem is the solution of the "Jacobian problem" that defines the deformation $\mathbf{f}$.

We say that $\mathbf{f}$ is a *periodic deformation* on the cube Ω with sidelenght L if $\mathbf{f}$ is a C^1 diffeomorphism on $\mathbf{R}^3$ leaving Ω invariant, i.e. $\mathbf{f}(\Omega) = \Omega$, and if $\mathbf{f}(\mathbf{x} + L\mathbf{m}) = \mathbf{f}(\mathbf{x}) + L\mathbf{m}$; for any $\mathbf{m} \in \mathbf{Z}^3$ and $\mathbf{x} \in \mathbf{R}^3$. This means that $\mathbf{f}$ is a C^1 diffeomorphism on the torus or its periodic extension to $\mathbf{R}^3$. We denote $J_{\mathbf{f}}(x) := \det(D\mathbf{f}(x))$ the Jacobian of $\mathbf{f}$.

In our context we want to transform the given constant density ρ_0 of plane waves (or of some wavefunction) into the actual x dependent density $\rho(x)$.

In order to derive energy formulas such as (16), the deformation method allows to construct a set of functions (ψ_j), with density $\sum_{j=1}^{N} |\psi_j(x)|^2 = \rho(x)$, satisfying orthonormality constraints $\langle \psi_i, \psi_j \rangle = \delta_{ij}$.

It is easy to see that the deformed plane waves defined by

$$\psi_{j,\mathbf{f}}(x) = \frac{1}{\sqrt{|\Omega|}} \, J_{\mathbf{f}}(x)^{1/2} \, e^{ik_j \cdot \mathbf{f}(x)} \tag{10}$$

where the deformation $\mathbf{f}$ is chosen as in Lemma 1 below and where the wave numbers, $k_j \in \frac{2\pi}{L}\mathbf{Z}^3$, are all distinct, are a solution to the above problem, as soon as the deformation $\mathbf{f}$ is a solution of the Jacobian problem $J_{\mathbf{f}}(x) = \rho(x)/\rho_0$. (Note that a change of variable will give the conservation of the orthonormality constraints.)

Based on the fundamental result of Dacorogna and Moser [12] we obtain the following

Lemma. 1 *There is an $\eta(\alpha) > 0$ and a $K(\alpha) > 0$ such that, $\forall \rho \in D_\alpha$ that satisfy $\|\epsilon_\rho\|_{0,\alpha} < \eta(\alpha)$, there exists a periodic deformation $\mathbf{f}$ that is solution of the Jacobian equation $J_{\mathbf{f}}(x) = \frac{\rho(x)}{\rho_0}$ such that*

$$\|\mathbf{f} - Id\|_\infty + \|D\mathbf{f} - Id\|_\infty + \|D\mathbf{f} - Id\|_{0,\alpha} \leq K(\alpha)\|\epsilon_\rho\|_{0,\alpha}. \tag{11}$$

Note that the Hölder continous assumption on ρ comes from the use of the above lemma. We believe it should be possible to use weaker assumptions in some Sobolev space.

3   Kinetic energy estimates

Our first result states that the exact kinetic energy, as a functional of the density, is equal to the Thomas-Fermi-von Weizsäcker functional [20] up to small remainder terms depending on ϵ_ρ and N.

Theorem. 1 *Let $0 < \alpha < 1$. For densities $\rho \in D_\alpha$, the functional $\mathcal{E}_{kin}(\rho)$ has the following asymptotic behaviour in a neighborhood of $\rho = \rho_0$ and $N = +\infty$:*

$$\mathcal{E}_{kin}(\rho) = \int_\Omega |\nabla\sqrt{\rho}|^2 dx + C_F \int_\Omega \rho^{5/3} dx \Big\{ 1 + O(\|\epsilon_\rho\|_{0,\alpha}) + O(\frac{1}{N^{1/2}}) \Big\} \tag{12}$$

where $C_F := \frac{3}{5}(3\pi^2)^{2/3}$ is the Fermi constant (in our context).
Furthermore there exists $\eta(\alpha) > 0$ such that the error terms are uniform with respect to pairs (N, ρ) satisfying $\|\epsilon_\rho\|_{0,\alpha} < \eta(\alpha)$.

Note that this Theorem remains true if we make another choice of a periodic Hamiltonian (and thus of the periodic model). Actually Theorem 1 depends only on the choice of the space Λ of wave functions. However, this result becomes physically relevant only if, for the choosen model, we are able to prove that the density of the ground state is closed to the constant density.

The "high density limit" is obtained by letting $N \to \infty$ for a given size of the periodicity cell. In this limit and for our choice of the periodic model, the assumption $\epsilon_\rho \to 0$, is physically reasonable. Nevertheless this assumption is not yet mathematically proved in a general context (see [19] for a proof in the Thomas-Fermi context and [19], [20],[15] for a discussion on different types of limits).

The other error term in (12), $O(\frac{1}{N^{1/2}})$, is due to the approximation of the kinetic energy of plane waves with wave vectors in a continous sphere by the kinetic energy for the N discrete wave vectors. This estimate on $\mathcal{E}_{kin}(\rho_0)$, that we did not find in the literature in this form, relies on results by Skriganov [27]. (Cf. also Remark 4 after Theorem 3 on the importance of a good asymptotic estimation of the kinetic energy).

The estimate (12) is also valid locally: denoting
$T_{loc}(\Psi)(x) := N \int_{\Omega^{N-1}} |\nabla \Psi(x, x_2, \ldots, x_N)|^2 dx_2 \cdots dx_N$ we have:

$$T_{loc}(\Psi)(x) = |\nabla \sqrt{\rho}(x)|^2 + C_F \rho^{5/3}(x)\{1 + O(\|\epsilon_\rho\|_{0,\alpha}) + O(\frac{1}{N^{1/2}})\} \,,$$

for the wave function $\Psi \in \Lambda$ minimizing $T(\rho)$.

Furthermore, if we consider only wave functions which are Slater determinants, i.e. $\Psi(x_1, ..., x_N) = \frac{1}{\sqrt{N!}} \det(\psi_j(x_i))$, with $\int_\Omega \psi_i \bar{\psi}_j = \delta_{ij}$ (using $\psi_j = \psi_{j,f}$ as in (10)), we obtain with the same assumption as in Theorem 1 an upper bound of order 2 in ϵ_ρ (cf. [8]):

$$\mathcal{E}_{kin}(\rho) \leq \int_\Omega |\nabla \sqrt{\rho}|^2 \, dx + C_F \int_\Omega \rho^{5/3} \, dx \left\{1 + O(\|\epsilon_\rho\|_{0,\alpha}^2) + O(\frac{1}{N^{1/2}})\right\}. \qquad (13)$$

On the other hand, the estimate (12) allows us to prove, in our specific context, a well known conjecture of March and Young (cf. [24] and also [20]). Namely we have the following Corollary:

Corollary. 1 *There exists $C(\alpha) > 0$ and $\eta(\alpha) > 0$ such that $\forall \rho \in D_\alpha$ et $\forall N > 0$ satisfying $\|\epsilon\|_{0,\alpha} < \eta(\alpha)$ we have*

$$T(\rho) \leq \int_\Omega |\nabla \sqrt{\rho}|^2 dx + C(\alpha) \int_\Omega \rho^{5/3} dx. \qquad (14)$$

Remark. 1 *An application of these techniques in the case of a system of N fermions in $\mathbf{R}^3$ faces the problem to find an equivalent of ρ_0 in the non periodic case. Note that in [3] we have proven the estimate*

$$\mathcal{E}_{kin}(\rho) \leq C N^{2/3} \int_{\mathbf{R}^3} |\nabla \sqrt{\rho}|^2 dx$$

by deforming $\mathbf{R}^3$ onto the unit cube.

Remark. 2 *We recently learned about a similar approach in [14], based, however, on more formal estimations.*

4 Justification of the $X\alpha$ method and exchange energy approximation

Our second result concerns the local exchange potential occuring in the $X\alpha$ method (cf [22]) as an approximation to the exchange potential V_{ex} of the Hartree-Fock model. We recall that the Hartree Fock Energy is defined by :

$$E^{HF} := \min \left\{ \langle H_{per}\Psi,\ \Psi \rangle,\ \Psi \in \Lambda,\ \Psi = \frac{1}{\sqrt{N!}} det(\psi_i(x_j)),\ \langle \psi_i, \psi_j \rangle = \delta_{ij} \right\} \quad (15)$$

Since the constraints $\langle \psi_i, \psi_j \rangle = \delta_{ij}$, $1 \leq i, j \leq N$ implies that $\|\Psi\|_{L^2} = 1$, we have $E_0 \leq E^{HF}$. Then, a classical computation [22] gives for the above determinants Ψ :

$$\langle H_{per}\Psi,\ \Psi \rangle = \sum_{j=1}^{N} \int_\Omega |\nabla \psi_j|^2 + \int_\Omega \rho V_{ext} + J(\rho) + E_{ex}(\psi_j) \quad (16)$$

where $\rho(x) := \sum_{j=1}^{N} |\psi_j(x)|^2$, $J(\rho) := \frac{1}{2}\int_{\Omega^2} \rho(x)\rho(y)G(x,y)dxdy$ is the "*direct Coulomb energy*", and where E_{ex}, the *exchange energy*, is defined by:

$$E_{ex}(\psi_j) = -\frac{1}{2}\int_{\Omega\times\Omega} |D(x,y)|^2 G(x-y)\, dx\, dy\ , \quad (17)$$

with the density matrix $D(x,y) := \sum_{j=1}^{N} \psi_i(x)\overline{\psi_j}(y)$. The minimization of (16) under the constraints $\langle \psi_i, \psi_j \rangle = \delta_{ij}$ gives the so-called "HF equations" (the existence of a minimum for (16), which is a difficult problem when posed on $\mathbf{R}^3$ [23], can be more easily proved here since the torus Ω is compact) :

$$-\Delta\psi_i + V_{ext}(x)\psi_i + \left(\frac{1}{|x|} * \rho\right)\psi_i + (V_{ex}\psi_i)(x) = \sum_{j=1}^{N} \epsilon_{ij}\psi_j \quad (18)$$

where (ϵ_{ij}) is a constant hermitian matrix and where V_{ex} is a non-local operator, defined by:

$$(V_{ex}\psi_j)(x) := -\int_\Omega D(x,y)G(x-y)\psi_j(y)\, dy. \quad (19)$$

There is no exact local expression for the complicated *exchange potential* V_{ex}. However, V_{ex} can be astonishingly well approximated by $-C_S\rho^{1/3}(x)$ (for some constant C_S) as proposed by Slater [26] and widely used under the name "$X\alpha$ method". We refer to [22] for a review of such approximations. A first rigorous derivation of this Slater approximation based on deformations of plane waves has been given in [9]. Following [26] we first approximate the exact exchange potential V_{ex} by the *average exchange potential* V_{av}:

$$V_{av}(x) := -\int_\Omega \frac{|D(x,y)|^2}{\rho(x)} G(x-y)\, dy \quad (20)$$

This formula comes from the "Slater averaging" of the HF exchange potential $(V_{ex}\psi_i)(x)$ by the weigthed densities of the i-th wavefunction : $\sum_{j=1}^{N}(V_{ex}\psi_i)(x)\frac{|\psi_i|^2}{\rho(x)}$. The advantage of V_{av} is that it can be used as a "local" approximation of (18) : $(V_{ex}.\psi_j)(x) \sim V_{av}(x)\psi_j(x)$. Furthermore, we can recover the exact exchange energy from V_{av}, since we have from (15), (20) :

$$E_{ex}(\psi_j) = \frac{1}{2}\int_{\Omega}\rho(x)V_{av}(x). \qquad (21)$$

We have the following asymptotic results

Theorem. 2 *Let $0 < \alpha < 1$. For densities $\rho \in X_{\alpha}$, in a neighborhood of $\rho = \rho_0$ and $N = +\infty$, and with the deformed plane waves (10), we have*
(i) Uniformly in $x \in \Omega$,

$$V_{av}(x) = -C_S\,\rho(x)^{1/3}\Big\{1 + O(\|\epsilon_\rho\|_{0,\alpha}^{1+\alpha}) + O(\frac{1}{N^{1/3}})\Big\}$$

(ii)

$$E_{ex}(\psi_j) = -C_S\int_{\Omega}\rho^{4/3}dx\Big\{1 + O(\|\epsilon_\rho\|_{0,\alpha}^{1+\alpha}) + O(\frac{1}{N^{1/3}})\Big\}$$

Note that (ii) is an immediate consequence of (i) and of (21). Statement (i) of Theorem 2 is proved in [8]. Basically, we follow the proof of [9], and use the work of Friesecke [15] in order to deal with planes waves whose wave numbers are in the Fermi sphere (instead of the cube used for simplicity in [24], [9]).

Note also that Theorem 2 gives a justification of the Dirac approximation $\int_{\Omega}\rho^{4/3}dx$ (see [13]) of the exchange energy.

5   Global energy bound

Our method of deformation also gives an upper bound for the energy $\mathcal{E}(\rho)$. Nevertheless, we have to restrict ourselves to wave functions of Slater determinant type which are deformations of plane waves (see (10)) in the same way as for the second order bound for the kinetic energy. This is why we only obtain an upper bound for the global energy. Combining the kinetic and exchange energy estimations, we obtain:

Theorem. 3 *Let $0 < \alpha < 1$. For densities $\rho \in D_{\alpha}$, the functional $\mathcal{E}(\rho)$ admits the following upper bound in a neighborhood of $\rho = \rho_0$ and $N = +\infty$:*

$$\mathcal{E}(\rho) \le \int_{\Omega}|\nabla\sqrt{\rho}|^2dx + C_F\int_{\Omega}\rho^{5/3}dx\Big\{1 + O(\|\epsilon_\rho\|_{0,\alpha}^2) + O(\frac{1}{N^{1/2}})\Big\}$$

$$+ \int_{\Omega}V_{ext}(x)\rho(x)dx + J(\rho) - C_S\int_{\Omega}\rho^{4/3}(x)dx \qquad (22)$$

where $J(\rho) = \frac{1}{2}\int_{\Omega^2}\rho(x)\rho(y)G(x,y)\,dxdy$ is the Coulomb energy, $C_F := \frac{3}{5}(3\pi^2)^{2/3}$ is the Fermi constant and $C_S := \frac{3}{2}(\frac{3}{\pi})^{1/3}$ is the "Slater" constant (in our context). Furthermore there exists $\eta(\alpha) > 0$ such that the error terms are uniform with respect to couples (N, ρ) satisfying $\|\epsilon_\rho\|_{0,\alpha} < \eta(\alpha)$.

Remark. 3 *The error terms in Theorem 3 can also be bounded as follows:*

$$\int_\Omega \rho^{5/3} \left\{ O(\|\epsilon_\rho\|_{0,\alpha}^2) + O(\frac{1}{N^{1/2}}) \right\} \le cst. \ N^{5/3} \left\{ \|\epsilon_\rho\|_{0,\alpha}^2 + \frac{1}{N^{1/2}} \right\} \tag{23}$$

Remark. 4 *When we compare Theorem 3 with our result in [6], the leading error term in the TFvW approximation of the kinetic energy is improved (we obtain $O(\frac{1}{N^{1/2}})$ instead of $O(\frac{1}{N^{1/3}})$ in [6]). This is technically much more complicated but essential for combining it with the Dirac term, $\int \rho^{4/3} \sim cst. \ N^{4/3}$ which is now asymptotically larger than the improved error term, $O(N^{5/3} \cdot \frac{1}{N^{1/2}}) = O(N^{7/6})$.*

Acknowledgements. Support by the TMR network "Asymptotic Methods in Kinetic Theory" is acknowledged. The third author (NJM) acknowledges financial support by the DAAD-PROCOPE and by his FWF "Erwin Schrödinger Fellowship". The first author (O.B.) also thanks I. Catto for signaling Ref. [15] of Friesecke.

Bibliography

[1] N.W.Ashcroft and N.D.Mermin *Solid State Physics* , Holt, Rinehart, Winston (1976)

[2] V. Bach "Accuracy of Mean Field Approximations for Atoms and Molecules" *Commun. Math. Phys.* **155** (1993) 295 - 310

[3] O. Bokanowski and B. Grébert "A decomposition theorem for wave functions in Molecular Quantum Chemistry" *Math. Models & Methods Appl. Sci.*, **6** (1996) 437 - 466.

[4] O. Bokanowski and B. Grébert "Deformations of density functions in molecular quantum chemistry" *J. Math. Phys.* **37** (1996) 1553 - 1573.

[5] O. Bokanowski and B. Grébert "Utilization of deformations in molecular quantum chemistry and application to Density-Functional Theory" *Int. J. Quantum Chem.* **68**(1998) 221 -231.

[6] O. Bokanowski, B. Grébert and N.J. Mauser, "Approximations de l'énergie cinétique en fonction de la densité pour un modèle de Coulomb périodique". *C.R.Acad.Sci., Math.Phys.* **329** (1999) 85–90

[7] O. Bokanowski, B. Grébert and N.J. Mauser, "Rigorous derivation of the "Xα" exchange potential: a deformation approach", to appear in *Theo.Chem.* (2000)

[8] O. Bokanowski, B. Grébert and N.J. Mauser, in preparation.

[9] O. Bokanowski and N.J. Mauser "Local approximation for the Hartree-Fock exchange potential: a deformation approach" *Math. Models & Methods Appl. Sci.*, Vol. 9, No 6 (1999), 941-961.

[10] I. Catto, C. Le Bris, and P.L. Lions "Limite thermodynamique pour les modèles de type Thomas-Fermi" *C.R.Acad.Sci.* **322** (1996) 357-364.

[11] I. Catto, C. Le Bris, and P.L. Lions "Sur la limite thermodynamique pour les modèles de type Hartree et Hartree-Fock" *C.R.Acad.Sci.* **327** (1998) 259 - 266.

[12] B. Dacorogna and J. Moser "On a partial differential equation involving the Jacobian determinant" *Ann. Inst. Henri Poincaré, Analyse non linéaire,* **7** (1990) 1-26

[13] P.A.M. Dirac "Note on exchange phenomena in the Thomas-Fermi atom" *Proc. Cambridge Philos. Soc.,* **26** (1931) 376-385

[14] L. De Santis and R. Resta, "N-Representability and density-functional construction in curvilinear coordinates" *Solid State Commun.* **106**, 763 (1998).

[15] G. Friesecke "Pair Correlation and Exchange Phenomena in the Free Electron gas" *Comm. Math. Phys.* **184** (1997) 143 - 171.

[16] G.M. Graf and J.P. Solovej "A correlation estimate with application to quantum systems with Coulomb interactions" *Rev.Math.Phys.,* **6**, 5a (1994) 977– 997

[17] E.S. Kryachko and E.V. Ludeña "Formulation of N- and v- representable density-functional theory. I. Groundstates" *Phys. Rev.* **A 43** (1991) 2179 - 2193

[18] E.H. Lieb "Density functional for Coulomb systems" *Int. J. Quantum Chem.* **24** (1983) 243.

[19] E.H. Lieb and B. Simon "The Thomas-Fermi Theory of Atoms, Molecules and Solids" *Adv. Math.* **23** (1977) 22 - 116

[20] E.H. Lieb "Thomas-Fermi and Related Theory of Atoms and Molecules" *Rev. Mod Phys.* **53** (1981) 603 - 641

[21] E.H. Lieb and B. Simon "The Hartree-Fock theory for Coulomb systems" *Comm. Math. Phys.* 53 (1977) 185 - 194

[22] R.G. Parr and W. Yang "Density functional theory of Atoms and Molecules" (Oxford university press, 1989).

[23] P.L. Lions "Solution of Hartree-Fock equations for Coulomb systems" *Comm. Math. Phys.* **109** (1987) 33 - 97

[24] N.H. March, W.H. Young, "Variational methods based on the Density Matrix" *Proc. Phys. Soc.* **72** (1958) 182-192.

[25] I. Petkov, M. Stoitsov and E. Kryachko "Method of local-scaling transformation and DFT in Quantum Chemistry, part I." *Int. J. Quantum. Chem.* **29** (1986) 149 - 161.

[26] S J.C. Slater "A simplification of the Hartree-Fock method" *Phys. Rev.* **81** (3) (1951) 385 - 390.

[27] M.M. Skriganov, "Geometric and Arithmetic Methods in the Spectral Theory of Multidimensional Periodic Operators" *Proc. Steklov Instit. of Math.* (1987) issue 2

[28] Dong Yé "Prescribing the Jacobian determinant in Sobolev spaces" *Ann. Inst. Poincaré, Analyse non linéaire* **11** (3) (1994) 275 - 296.

Chapter 7

A mathematical insight into *ab initio* simulations of the solid phase

X. Blanc
Ecole Normale Supérieure
45 rue d'Ulm, 75230 Paris Cedex 5, France.
xblanc@ens.fr

Abstract: We give here an introduction to the language of *ab initio* solid-state theory. Starting from the intrinseque symmetries of a perfect crystal, we show how it is related to band structure theory and give a brief overview of the techniques in use to simulate such systems.

1   Introduction

This article is concerned with *ab initio* numerical solid-state chemistry, that is, the computation of electronic structure in a solid. We will only consider the case of perfect crystals, i.e an infinite periodic arrangement of nuclei, in which electrons are supposed to satisfy the Schrödinger equation. Of course, some extensions are possible, in particular to treat defects in solids (see [71], chapters 12 and 13 for instance).

We start with some definitions, introducing the standard language of lattice structure. Good complements to this introduction may be found in [2, 51]. See also [32] (or any crystallographic textbook) for a more theoretical work. Some extensions to quasi-crystals may be found in [48, 74] for instance.

Section 3 is devoted to the language of band theory, which accounts for the distinction between metals and insulators, and is based on the Bloch theorem. Here again, standard solid-state textbooks, such as [2, 51] provide a very good overview on the topic. Concerning the corresponding mathematical problems, we refer to [73, 8, 55, 29].

Section 4 presents standard *ab initio* theories, starting from the molecular formulation (which means that it involves only a finite number of nuclei and electrons, [66, 67]), and extending it to the case of an infinite periodic crystal. These theories are presented in [71] and [65]. We also give many references in the course of this section, since many problems and methods are overlooked here.

The last section is devoted to the practical aspects of these theories, pointing out the most important numerical features and limitations.

2   Crystalline structure

We consider a perfect crystal, that is, a system of infinite size consisting of a periodically repeated set of nuclei, together with an associated set of electrons. Assuming that the nuclei are fixed, we may describe their positions by a motif m repeated on the nodes of a lattice $\mathcal{R}$:

$$\mathcal{R} = \{ua + vb + wc, \quad u, v, w \in \mathbf{Z}\}. \tag{1}$$

The vectors a, b, c are assumed to be linearly independent vectors of $\mathbf{R}^3$, and are called a basis of $\mathcal{R}$. Using the approximation of classical nuclei, the motif m represents N point particles of positive charge, that is, mathematically speaking, a set of positive Dirac masses :

$$m = \sum_{i=1}^{N} Z_i \delta_{x_i}, \tag{2}$$

where the vectors x_i may be chosen to be in the set :

$$\Gamma(a, b, c) = \{xa + yb + zc, \quad x, y, z \in [-\frac{1}{2}, \frac{1}{2}[\} \tag{3}$$

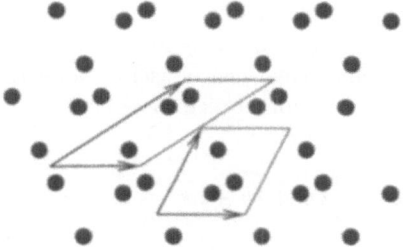

Figure 7.1: The set of nuclei with two possible unit cells and lattice basis

We call a *unit cell* of the lattice $\mathcal{R}$ any semi-open polyhedron such that its translations along a subset of the vectors of the lattice fills in the whole space $\mathbf{R}^3$ without overlapping. The set $\Gamma(a, b, c)$ defined in (3) is thus a unit cell of $\mathcal{R}$. Many choices of unit cells are possible. If its volume is minimal, which is the case of $\Gamma(a, b, c)$, we call the cell a *primitive* unit cell. Such a cell satisfies the property that, when translated by *all* the vectors of the lattice, it constitutes a partition of $\mathbf{R}^3$. A particular primitive unit cell is the Wigner-Seitz cell, or Dirichlet zone, which is constructed as follows :

(i) The point $0 \in \mathcal{R}$ is jointed to all of its neighbors by segments;

(ii) the mediator planes of these segments are traced;

(iii) each of these planes define a half-space containing 0, and the intersection of these half-spaces is a convex polyhedron, which is the Wigner-Seitz cell.

This cell presents the advantage of being unique.

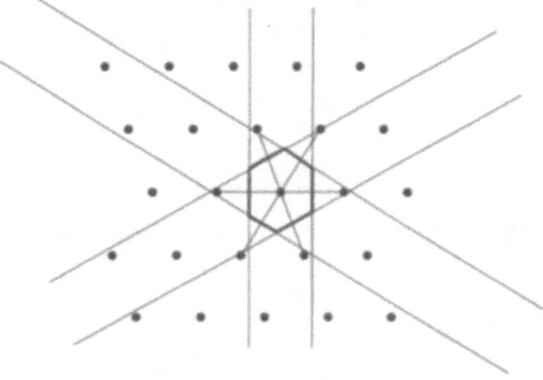

Figure 7.2: The Wigner-Seitz cell of a lattice

The aim of *ab initio* calculations is to describe the electronic structure of a crystal, the lattice on which the nuclei are set being given. These calculations handle many $\mathcal{R}$-periodic functions, for which treatment a useful tool is the Fourier analysis. This is why we now introduce the notion of reciprocal lattice.

Given a basis (a, b, c) of $\mathcal{R}$, we set :

$$\mathcal{R}^* = \left\{ ua' + vb' + cw', \quad u, v, w \in \mathbf{Z} \right\}, \tag{4}$$

where the basis (a', b', c') is defined as follows :

$$\begin{cases} a \cdot a' = b \cdot b' = c \cdot c' = 2\pi, \\ a \cdot b' = a \cdot c' = b \cdot c' = 0. \end{cases} \tag{5}$$

In other words, the matrices of columns respectively (a, b, c) and $\dfrac{1}{2\pi}(a', b', c')$ are inverse of each other. The point is that this definition of $\mathcal{R}^*$ does not depend on the chosen basis. The Wigner-Seitz cell of the reciprocal lattice is called the first Brillouin zone (henceforth abbreviated as BZ) of the crystal.

In many cases, the relevant parameter of the lattice is its invariance group, rather than the lattice itself, especially for optic properties. For definitions of these terms and discussions of related topics, we refer to [51, 2], where one will also find details on the above few assertions, and to [71], chapter 1, for a brief introduction.

3   Band theory

3.1   Band structure

It is a commonly accepted approximation that an electron of a solid experiences a periodic potential. This is why the first step in studying the electronic structure of a solid is the examination of the properties of a Schrödinger operator with periodic potential: let $H = -\Delta + V$ be such an operator, V being $\mathcal{R}$-periodic. The question is, can we find a spectral decomposition of H? This leads us to the following eigenvalue problem, or equivalently to the determination of stationary states of H:

$$-\Delta\psi + V\psi = E\psi. \tag{6}$$

Let us denote by τ_T the operator

$$\psi \longmapsto \psi(\cdot + T), \tag{7}$$

for any $T \in \mathcal{R}$. The point is that H is invariant under such translations, which means, in terms of operators:

$$\tau_T H = H\tau_T, \tag{8}$$

for all $T \in \mathcal{R}$. Since these two operators commute, they have a common diagonalizing basis. In other words, one can find a basis such that each of its element satisfies:

$$\begin{cases} H\psi = E\psi, \\ \tau_T\psi = \gamma(T)\psi, \quad \text{for some} \quad \gamma(T) \in \mathbf{C}. \end{cases} \tag{9}$$

Noticing that $\tau_{T+T'} = \tau_T \tau_{T'} = \tau_{T'} \tau_T$, we infer from (9) that $\gamma(T+T') = \gamma(T)\gamma(T')$. On the other hand, since τ_T is norm-preserving, we necessarily have $|\gamma(T)| = 1$. It follows that there exists a vector k such that:

$$\gamma(T) = e^{ik \cdot T}. \tag{10}$$

From the fact that $T \in \mathcal{R}$, the addition of a vector of the reciprocal lattice $\mathcal{R}^*$ to k does not change $\gamma(T)$, so that we may impose that k is in some primitive unit cell of $\mathcal{R}^*$, for example the first Brillouin zone (BZ). Equation (9) thus becomes:

$$\begin{cases} H\psi = E\psi, \\ \psi(x+T) = e^{ik \cdot T}\psi(x), \quad \text{with} \quad k \in BZ. \end{cases} \tag{11}$$

This leads us to the celebrated Bloch theorem:

Theorem 3.1 *Let $H = -\Delta + V$ be a Schrödinger operator, with $V \in L^2_{loc}$ $\mathcal{R}$-periodic. Then for all $k \in BZ$, there exists an increasing sequence of eigenvalues $\{\lambda_j(k)\}$ and a corresponding sequence of eigenvectors $\{\psi_j^k\}$ of H such that :*

- $k \mapsto \lambda_j(k)$ *is continuous;*

- $\psi_j^k(x)e^{-ik \cdot x}$ *is periodic;*

- $\sigma(H) = \bigcup_{j \in \mathbf{N}} [\inf_{BZ} \lambda_j, \sup_{BZ} \lambda_j].$

($\sigma(H)$ denotes the spectrum of the operator H.)

The above argument is by no means a proof of this result. The interested reader may find more detailed proofs in [73, 55, 29]. The original idea of decomposing the spectrum of this type of Schrödinger operators by looking for solutions which are periodic up to k is due to Bloch [13].

The Bloch theorem gives a spectral decomposition of any Schrödinger operator with periodic potential. In particular, we see that the spectrum of such an operator is constituted of intervals, which are called *bands* of the operator. This gives us an intuition on the behavior of electrons in a crystal: assuming that they are represented by the above eigenstates, preferably the lowest ones, we see that if there are exactly enough electrons to fill in an entire number of bands, we have an energetical gap between the highest non-excited state and the lowest excited one. Conversely, if a band is only partially filled, there is no such a gap. In the first case, one needs a great amount of energy to take an electron from its stationary state to an excited state, and the crystal behaves like an insulator. In the second case, a small amount of energy is sufficient for an electron to reach an excited state: this is a metallic behavior. This is shown in Figure 7.3.

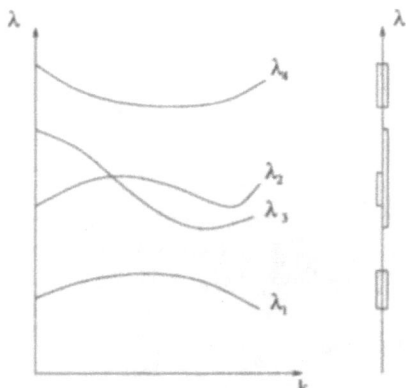

Figure 7.3: The band structure of a one-dimensional crystal: if there is one electron per cell, only the first band is filled and the crystal is an insulator, whereas if there are two electrons per cell, the second band is only partially filled, so that we have a metal.

3.2   Density of states

If one simulates a crystal at zero-temperature, the above Bloch states are filled in using the *Aufbau* principle, that is, starting from the lowest energy state, then the next lowest, and so on. Of course, this is a much clearer concept in the molecular case, where the are only a numerable quantity of eigenstates, than in the crystal case, where the energy levels are continuously ordered. In this latter case, we know that we have a fixed number of electrons per unit cell, although an electron associated to a precise unit cell need not be localized in this cell. Thus, one eigenstate does not represent an electron by itself, but the set of all similar electrons repeated through the translations by the lattice vectors (physically, this corresponds to delocalized electrons).

Mathematically, this corresponds to defining the following function:

$$n(\lambda) = \frac{2}{|BZ|} \sum_{j \in \mathbf{N}} \int_{BZ} \delta(\lambda - \lambda_j(k)) dk, \tag{12}$$

where $|BZ|$ is the volume of the BZ, and δ a Dirac mass at 0. The function n is called the density of state, and the function f defined by

$$f(\varepsilon) = \int_{-\infty}^{\varepsilon} n(\lambda) d\lambda \tag{13}$$

represents the number of electron states per unit cell with energy lower than ε. The factor 2 in (12) accounts for the fact that we use the closed-shell approximation.[1]

---

[1]The closed-shell approximation corresponds to neglecting the spin of the electrons, and considering two electrons with the same wave function and opposite spin as one particle.

The function f is non-decreasing, since $n(\lambda) \geq 0$, and we define the Fermi energy ε_F by

$$f(\varepsilon_F) = N, \tag{14}$$

where N is the number of electrons per cell. By doing this, we have filled in entirely the levels lower than the Fermi energy, leaving the higher energy states empty.

In general, when one computes a macroscopic parameter of a crystal, it often appears in the form:

$$S = \frac{2}{|BZ|} \sum_j \int_{BZ} g(j, k)dk, \tag{15}$$

where g is a function of the electronic state labeled by j and k. In many cases, the function g depends only on the energy: $g(j, k) = g(\lambda_j(k))$. Thus, using the definition (12) of n and changing variables in the above definition of S, we have:

$$S = \int_{-\infty}^{\varepsilon_F} g(\lambda)n(\lambda)d\lambda. \tag{16}$$

The density of states thus appears as an important feature of the crystal, in the sense that its values determine those of many of the crystal's macroscopic parameters.

In the case where the temperature T is not 0, one only needs to replace the Heaviside distribution implicitly used above by the Fermi-Dirac distribution $\mu_T(x) = \frac{1}{1+e^{-x/k_BT}}$. Equation (12) would thus become:

$$
\begin{aligned}
n_T(\lambda) &= \frac{2}{|BZ|} \sum_{j\in\mathbf{N}} \int_{BZ} \mu_T'(\lambda - \lambda_j(k))dk \\
&= \frac{2}{|BZ|} \sum_{j\in\mathbf{N}} \int_{BZ} \frac{1}{k_BT} \frac{e^{-\frac{\lambda-\lambda_j(k)}{k_BT}}}{1 + e^{-\frac{\lambda-\lambda_j(k)}{k_BT}}} dk.
\end{aligned}
\tag{17}
$$

So that f becomes

$$f_T(\varepsilon) = \frac{2}{|BZ|} \sum_{j\in\mathbf{N}} \int_{BZ} \mu_T(\varepsilon - \lambda_j(k))dk. \tag{18}$$

Note that in practice, the above integrals over the BZ are replaced by sums over some points of the BZ, according to a prescribed sampling. The most natural sampling is the regular one, in which one distributes M^3 points regularly in the BZ. This in fact corresponds to a periodic boundary condition on a box composed of M^3 unit cells. Indeed, in this case, boundary conditions impose that k takes a finite number of values, namely $(\frac{k_1}{M}a, \frac{k_2}{M}b, \frac{k_3}{M}c)$, with $0 \leq k_i < M$. However, and although this is appealing to justify periodic boundary conditions, this sampling is not optimal, and special points technics [33, 16, 63] need to be used, extensively using the symmetries of the crystal.

4   Electronic calculations: standard approximations

Let us now consider the numerical *ab initio* computations of electronic structure in a crystal, and more specifically the approximations commonly used. So far, we have dealt with the Schrödinger equation satisfied by *one* electron in a periodic potential. Although this is a qualitatively good model, computations have to be more precise.

The approximation of classical nuclei, mentioned in section 2, and which consists in considering that the nuclei are classical point particles, is widely used. It is justified by the fact that the mass of a nucleus is much larger than that of an electron. In the same spirit we will also use the Born Oppenheimer approximation, that is, the nuclei will be supposed to be fixed at their positions, the reason for this being that the electrons move much faster than the nuclei, and their movement will be treated in a second step, considering the electronic energy as an external potential for their movement. Concerning these lattice dynamics treatments, we refer to [71], Chapter 12 and to the references therein.

Another approximation currently made (and used so far) is the non-relativistic treatment of the electrons. This may be a poor one in some cases, especially in heavy nuclei where the speed of the core electrons is of the same order as the velocity of light c ($\frac{v}{c} \simeq 0.58$ in the case of Mercury). But this may be taken into account through relativistic pseudo-potentials (see below), or by an *a posteriori* perturbative treatment (see [71]).

Even with these approximations, solving Schrödinger equation is out of reach, in particular because the wave function of the electrons should be a function of NM spatial variables, where N is the number of electrons per cell and M the number of cells. Furthermore, we should rigorously take the limit $M \to \infty$. We thus need to either approximate this wave function (this the case of Hartree-Fock theory, see section 4.1), or use a different way to characterize the electrons state than using their wave function, which is the aim of density functional theory (DFT), reviewed in Section 4.2.

We only give here the equations in a formal way, assuming that they take the same form as in the case of a finite molecule, allowing to state in a more simple way the corresponding theories. We will then point out the difficulties conveyed by the fact that we are studying a solid, particularly the infinity of the number of particles (electrons as well as nuclei), and the periodicity of the system.

4.1   Hartree-Fock formalism

The Hartree-Fock (HF) theory of molecules aims at solving the electronic Schrödinger equation under the assumption that the wave function has the special form of a Slater determinant. It is a way of taking the exclusion principle into account. In this case of a finite number of electrons, say n, the total wave function reads $\phi(x_1, x_2, \ldots, x_n) = \det(\phi_i(x_j))$, and this gives the well-known Hartree-Fock equations:

$$-\frac{1}{2}\Delta\phi_i + V\phi_i + (\rho \star \frac{1}{|x|})\phi_i - \int_{\mathbf{R}^3} \frac{\tau(x,x')}{|x-x'|}\phi_i(x')dx' = -\varepsilon_i\phi_i, \tag{19}$$

with $\tau(x,x') = \sum_{i=1}^n \phi_i(x)\phi_i(x')^*$, $\rho(x) = \tau(x,x)$, and V represents the electrostatic potential created by the nuclei[2]; $\star$ denotes the convolution product over $\mathbf{R}^3$.

$$\begin{aligned} E^{HF}(\{\phi_i\}) &= \frac{1}{2}\sum_{i=1}^n \int_{\mathbf{R}^3} |\nabla\phi_i|^2 + \int_{\mathbf{R}^3} \rho V + \frac{1}{2}\int_{\mathbf{R}^3}\int_{\mathbf{R}^3} \frac{\rho(x)\rho(x')}{|x-x'|}dxdx' \\ &\quad - \frac{1}{2}\int_{\mathbf{R}^3}\int_{\mathbf{R}^3} \frac{|\tau(x,x')|^2}{|x-x'|}dxdx' + \frac{1}{2}\sum_{j\neq i} \frac{1}{|X_i - X_j|}, \end{aligned} \tag{20}$$

where the last term accounts for the repulsion of the nuclei, and is added so that the total charge of the system becomes 0, which allows to define its electrostatic interaction (see section 5.3). We see here that although these equations clearly make sense in the case where n is finite, the solid-state case is *not that obvious*. See [20] for the related mathematical problems. Nevertheless, equation (19) does make sense in this case, as soon as the total potential $V_{tot} = V + \rho \star \frac{1}{|x|}$ is not splitted into these two terms, but rather kept together so as to define the potential of a periodic neutral distribution of charge, which does exist.

The Fock hamiltonian may be written in the following form:

$$F = -\frac{1}{2}\Delta + V + \rho \star \frac{1}{|x|} - \sum_i \int_{\mathbf{R}^3} \frac{\phi_i(x')^* \bullet}{|x-x'|}\phi_i(x)dx', \tag{21}$$

where $V_{tot} = V + \rho \star \frac{1}{|x|}$ is to be understood as a solution of the Poisson equation:

$$-\Delta V_{tot} = 4\pi\left(\sum_{j\in\mathcal{R}} m(\cdot + j) - \rho\right), \tag{22}$$

the measure m defining the nuclei in the primitive unit cell (2). This equation has a periodic solution as soon as $m - \rho$ is neutral over the unit cell of $\mathcal{R}$. A similar treatment is possible for the exchange term (last term in (21)), see [20]. Equation (19) thus reads:

$$F\phi_i = -\varepsilon_i\phi_i. \tag{23}$$

In the case of a solid, the above equations are derived through Bloch's theorem, so that all sums over i are replace by sums over i *and* integrals over the BZ, and the wave functions and energies are labelled by $k \in BZ$. In other words, setting

$$\tau(x,x') = \sum_{j\in\mathbf{N}} \int_{BZ} \phi_j^k(x)\phi_j^{k*}(x')(\varepsilon_F - \varepsilon_j^k)^+ dk, \tag{24}$$

where $t^+ = \max(t,0)$, i.e the term $(\varepsilon_F - \varepsilon_j^k)^+$ selects only the states having lower energy than the Fermi energy ε_F defined in (14), and F being defined by

$$F\phi = -\frac{1}{2}\Delta\phi + V_{tot}\phi - \int_{\mathbf{R}^3} \frac{\tau(x,x')}{|x-x'|}\phi(x')dx', \tag{25}$$

---

[2]$V(x) = -\sum_{j=1}^p \frac{Z_j}{|x-X_j|}$, where p is the number of nuclei, and X_j and Z_j respectively their positions and charges.

with V_{tot} defined by (22), the ϕ_i^k should satisfy

$$F\phi_i^k = -\varepsilon_i^k \phi_i^k. \tag{26}$$

4.2   Density functional theory

Density functional theory (DFT) takes the problem from another standpoint than the one used by HF theory: instead of approximating the wave function, it approximates the total energy, writing it as a function of the electronic density ρ. Its theoretical ground is a theorem by Hohenberg and Kohn [49], which asserts that there exists a functional F of the density, independent of the external potential V, such that the electronic density of the ground state of the n-electrons problem with external potential V is a solution of:

$$\inf\left\{ F(\rho) + \int_{\mathbf{R}^3} \rho V, \quad \rho \geq 0, \quad \int_{\mathbf{R}^3} \rho = n \right\}. \tag{27}$$

Kohn and Sham proposed a computational framework in relation with this theorem in [53]. It consists essentially in giving a guess on the functional F. The choice of Kohn and Sham corresponds to:

$$F(\rho) = T(\rho) + \frac{1}{2} \int_{\mathbf{R}^3} \int_{\mathbf{R}^3} \frac{\rho(x)\rho(x')}{|x - x'|} dx dx' + E_x(\rho) + E_c(\rho), \tag{28}$$

where

$$T(\rho) = \inf\left\{ \frac{1}{2} \sum_{i=1}^n \int_{\mathbf{R}^3} |\nabla \phi_i|^2, \quad \sum_{i=1}^n |\phi_i|^2 = \rho \right\}, \tag{29}$$

and E_x and E_c are respectively the exchange and correlation energies. The local density approximation (LDA), currently used, consists in using the exchange energy of a homogeneous electron gas, that is:

$$E_x^{LDA}(\rho) = -\frac{3}{4}\left(\frac{3}{\pi}\right)^{1/3} \int_{\mathbf{R}^3} \rho^{4/3}. \tag{30}$$

Perdew and Zunger proposed in [70] for the correlation energy

$$E_c^{PZ}(\rho) = \int_{\mathbf{R}^3} \rho \epsilon_c\left(\left(\frac{3}{4\pi\rho}\right)^{1/3} \right), \tag{31}$$

with

$$\epsilon_c(r) = \begin{cases} 0.00311 \log(r) - 0.0480 + 0.0020r \log(r) - 0.0116r & \text{if } r < 1, \\ -0.1423/(1 + 1.0529\sqrt{r} + 0.3334r) & \text{if } r \geq 1. \end{cases} \tag{32}$$

Many other choices are possible, for example the $X\alpha$ setting [22], where $E_x + E_c = \alpha \int_{\mathbf{R}^3} \rho^{4/3}$, or the generalized gradient approximations (GGA), in which corrections to LDA are brought by terms of the form $G(\nabla\rho, \nabla^2\rho)$. See [10, 56, 69, 67] for instance.

In practice, this way of computing the density is used in the following manner:

$$\begin{cases} -\Delta\phi_i + V_{\text{eff}}\phi_i = -\varepsilon_i\phi_i, & \forall i, \\ V_{\text{eff}} = V_{\text{nuclei}} + \rho \star \frac{1}{|x|} + \frac{d}{d\rho}E_x + \frac{d}{d\rho}E_c, \\ \rho = \sum_i |\phi_i|^2, \end{cases} \tag{33}$$

where the ε_i are the lowest n eigenvalues of the operator $-\Delta + V_{\text{eff}}$. In the case of a solid, one needs to group the electrostatic terms together so as to be able to compute it through Poisson's equation (22). Moreover, the same dependance on the wave vector k as in HF theory appears here. Therefore, (33) becomes:

$$\begin{cases} -\Delta\phi_i^k + V_{\text{eff}}\phi_i^k = -\varepsilon_i^k\phi_i^k, & \forall i, k \\ V_{\text{eff}} = V_{tot} + \frac{d}{d\rho}E_x + \frac{d}{d\rho}E_c, \\ \rho = \sum_i \int_{BZ} |\phi_i^k|^2(\varepsilon_F - \varepsilon_j^k)^+ dk. \end{cases} \tag{34}$$

4.3 The use of pseudo-potentials

In many cases, the core electrons are physically almost unaffected by any change of environment. (This is particularly true in the case of low-energy transformations.) Hence, a natural approximation would be to consider them as frozen in some determined state (coming from the computation of a relevent atomic system), and compute only valence electrons. In order to implement this approximation, one introduces pseudo-potentials (PP). We consider here only their application to DFT, although a similar method carries through HF theory.

Consider a case where a number of (core) electrons are modelized by a pseudo-potential V_{ps}. Then, (34) would become:

$$\begin{cases} -\Delta\phi_i'^k + V_{\text{eff}}\phi_i'^k = -\varepsilon_i'^k\phi_i'^k, & \forall i, k \\ V_{\text{eff}} = V_{tot}' + \frac{d}{d\rho'}E_x + \frac{d}{d\rho'}E_c + V_{ps}, \\ \rho' = \sum_i \int_{BZ} |\phi_i'^k|^2(\varepsilon_F - \varepsilon_j^k)^+ dk, \end{cases} \tag{35}$$

where V_{tot}' accounts for the interaction between core electrons and the remaining electrons, which wave functions are denoted by $\phi_i'^k$. The PP operator must reproduce screened nuclear attractions, and account for Pauli exclusion principle, forcing the valence orbitals to be orthogonal to core ones. Moreover, the eigenvalues $\varepsilon_i'^k$ should be equal to the corresponding ε_i^k, and the corresponding eigenfunctions as close as possible to the "real" orbitals ϕ_i^k. And in the case of heavy atoms, the relativistic feature of the core electrons should be accounted for in V_{ps}. Above all this, the main feature of V_{ps} should be to simplify the computations, smoothing the Coulombic singalurity of the nucleus and having a usable expression. This gives a set of rules to elaborate PPs.

There exists a wide variety of pseudo-potentials, that we cannot display here. See for instance [6, 43, 76], concerning DFT pseudo-potentials, and [45, 46, 47] in the case of HF theory.

4.4   Self-consistent field algorithms

Solving the Hartree-Fock or the Kohn-Sham equations is done through the self-consistent field (SCF) procedure. It may be summarized as follows:

(i) Guess a first density $\rho = \rho_0$ (in the case of Kohn-Sham formalism), or a set of wave functions ϕ_i^k (in the case of Hartree-Fock setting), for each $k \in BZ$.

(ii) Compute the effective potential V_{eff}:

- in the KS formalism, $V_{\text{eff}} = V + \rho \star \frac{1}{|x|} + \frac{d}{d\rho}E_x + \frac{d}{d\rho}E_c$,[3]

- in the HF formalism, $V_{\text{eff}} = V + \rho \star \frac{1}{|x|} + \sum_j \int_{BZ}((\phi_j^k \bullet) \star \frac{1}{|x|}\phi_j^k)dk$.[3]

(iii) Compute eigenvalues $-\varepsilon_i^k$ and eigenvectors ϕ_i^k of $H = -\Delta + V_{\text{eff}}$, for each $k \in BZ$, and the Fermi energy ε_F.

(iv) Compute the new density $\rho = \sum_{i=1}^{N} \int_{BZ} |\phi_i^k|^2(\varepsilon_F - \varepsilon_j^k)^+ dk$.

(v) If the new ϕ_i^k's (in HF theory) or the new ρ (in KS theory) are equal to the preceding one(s), self-consistency is reached. Otherwise, go back to step (ii).

   In fact, important convergence problems occur in such an algorithm, and relaxation techniques need to be used to ease the convergence. See [19] for precisions.[4] In particular, when dealing with a metal, the levels near the Fermi surface may very well be alternatively filled and empty, generating oscillations. In this case, one needs, in a smoothing procedure, to partially fill each of them, which corresponds to the physical fact that these orbital behave as a mixing of themselves rather than individually. Another problem is that rigorously, one should at each SCF cycle compute the eigenvalues and eigenvectors $\varepsilon_i^k, \phi_i^k$ for all $k \in BZ$, which means numerically for many of them. In order to overcome this difficulty, one needs to choose these points very cleverly. This is done by special point technics (see Section 5.2 and [71], Chapters 4 and 10, or [63, 33, 62]). These technics usually allow to perform the calculation with only one (in the case of an insulator) to about twenty (in the case of a metal) k-points.

   From the fact that the algorithms used are the same in both HF and DF theory, one would think that they are quite similar technics. On the other hand, fundamental differences should be highlighted:

- The use of Kohn-Sham scheme yields an approximation of the kinetic energy that is not present in HF theory.

---

[3]V denotes the potential generated by the nuclei.

[4]The author is concerned with molecular models there, but convergence problems are in fact exactly the same as in solid-state models, as far as insulators are concerned. The case of metals is much more tricky, because of the Fermi surface.

- In the DFT formalism, the ε_i^k and ϕ_i^k obtained through the SCF algorithm have nothing to do *a priori* with the real electronic energies and orbitals. On the contrary, in the HF theory, they are interpretable in such a way. Furthermore, even the empty eigenstates of HF theory can be viewed in some way as excited states, according to Koopmans' theorem (see [66]).

- On the other hand, the HF theory neglects correlation effects, assuming that each electron interacts with the average field of the others. See [39] and [71], chapter 11 for a discussion of this topic. This problem accounts for large disagreement between HF theory and experiment, since correlation effects are usually important in solids. This may be dealt with through correlation-only energy [68, 39, 21, 3, 4].

- In HF theory, the exact treatment of the exchange energy implies more difficult computations. However, the simplification used in DFT raises inaccuracies. Furthermore, in DFT, the self-interaction of the electrons (second term in (28)) is overestimated since it contains an interaction of each electron with itself. Hence, there is a need for a self-interaction correction (SIC). Several such corrections have been proposed, which roughly consists in substracting from the energy the self-interaction of the computed pseudo-orbitals ϕ_i^k [50, 11, 70].

5   Numerical schemes

Let us now turn to the numerical schemes corresponding to the above theoretical framework. In both DFT and HF formalism, one chooses a basis (χ_j^k) of Bloch functions on which the crystalline orbitals of the electrons are developed, and solves the equations (26) or (34) in this basis, which become:

$$H^k C^k = S^k C^k E^k, \tag{36}$$

where H^k is the matrix of the corresponding Hamiltonian in the considered basis, S^k the overlap matrix, C^k the matrix of the crystalline orbitals and E^k is the diagonal matrix of the energies ε_i^k. In other words, we have:

$$\phi_i^k(x) = \sum_{j=1}^{N} c_{ji}^k \chi_j^k(x), \tag{37}$$

where each χ_j^k is periodic up to k, i.e satisfies the second equation of (11).

$$H_{ij}^k = \langle \chi_i^k | H | \chi_j^k \rangle, \quad S_{ij}^k = \langle \chi_i^k | \chi_j^k \rangle. \tag{38}$$

5.1   Basis sets

Hence, the first crucial choice is the Bloch functions χ_j^k. Let us briefly summarize the most widely used ones:

- Plane waves (PW): they present the advantage of carrying periodicity by themselves, and to be an orthogonal basis. Furthermore, the operator $-\Delta$ is diagonal in such a basis. Nevertheless, crystalline orbitals are usually not fairly regular near the nuclei, presenting cusps. Thus, their Fourier coefficients are not fastly decaying, so that one needs a great amount of plane waves to approximate in a good manner those orbitals.

- Gaussian-type orbitals (GTO): this corresponds to the choice

$$\chi_j^k(x) = \sum_{l \in \mathcal{R}} \chi_j(x - l)e^{ik \cdot l}, \tag{39}$$

where χ_j is a linear combination of Gaussians multiplied by spherical harmonics. They come from the molecular *ab initio* calculations, in which they are widely used because they make the computation of bielectronic integrals easier, using formulas due to Boys [17]. The point is, those molecular basis cannot be used directly in solid-state computations. In particular, very diffuse Gaussians increase dramatically the number of integrals to be explicitly calculated in the computation of electrostatic interactions.

In the HF formalism, these kinds of basis facilitate the computation of the exchange term, in the same fashion as in the molecular case [17].

Another good point concerning GTO is the fact that the Fourier transform of a Gaussian is a Gaussian. Thus, the completion of a GTO basis with PWs is computationally convenient [9]. Indeed, the additional problems raised by the computation of overlap and bielectronic integrals[5] are solved by the fact that (in the case of a PW and a GTO) the first ones are simply the Fourier transforms of Gaussians, and the second ones the Fourier transforms of Gaussians multiplied by the Fourier transform of $\frac{1}{|x|}$, that is, $\frac{1}{|\xi|^2}$. In this case, GTO functions are used to represent core electrons, which are localized near the nuclei, and valence electrons are represented by a mixed combination of PWs and GTOs.

- Muffin-tin orbitals (MTO): this method is a mixing of the two previous ones. The space is separated into two regions: the first one is a union of (non intersecting) spheres around the nuclei, which are called atomic spheres, and the second one is the remaining region, called the interstitial region. The muffin-tin approximation consists essentially in assuming that the potential and electron density are radially symmetric in the atomic spheres, and using PW expansion in the interstitial region. There exists a wide variety of such methods (APW, LAPW, FLAPW), for which we refer to [1, 35, 12, 5]. The FLAPW is also detailed in [71], chapter 9. As an example of such methods, we describe here the APW method:

As pointed out above, the main approximation of APW method relies on the separation of the space into MT spheres (one around each nucleus), and the remaining

---

[5] Of the form, respectively: $\int \chi_i^k \chi_j^k$ and $\int \int \frac{\chi_i^k(x)\chi_j^k(y)}{|x-y|}dxdy$.

region, which is called the interstitial region. In the MT spheres, the basis is a combination of solutions of radial Schrödinger equation:

$$\chi_j^k(x) = \sum_{l=0}^{\infty} \sum_{m=-l}^{l} A_{lm}(k + R_j) u_l(x - X_s) Y_{lm}\left(\frac{x - X_s}{|x - X_s|}\right), \tag{40}$$

where $R_j \in \mathcal{R}$, X_s is the position of the center of the MT sphere,[6] and Y_{lm} are the spherical harmonics. The function u_l is a solution of the radial Schrödinger equation:

$$-\frac{1}{r^2}\frac{d}{dr}\left(r^2\frac{du_l}{dr}\right) + \left[\frac{l(l+1)}{r^2} + V(r)\right]u_l = Eu_l. \tag{41}$$

In the interstitial region, the potential is assumed to be constant, and the wave functions are developped on a plane wave basis:

$$\chi_j^k(x) = e^{i(k+R_j)\cdot x}, \quad \text{with } R_j \in \mathcal{R}. \tag{42}$$

In order to have a basis of continuous functions, we need to equate the two values of χ_j^k on the MT spheres. Developping the exponential function on spherical harmonics, one gets:

$$e^{i(k+R_j)\cdot x} = 4\pi \sum_{l=0}^{\infty} \sum_{m=-l}^{l} i^l j_l(|k+R_j||x-X_s|) Y_{lm}\left(\frac{k+R_j}{|k+R_j|}\right)^* Y_{lm}\left(\frac{x-X_s}{|x-X_s|}\right) e^{i(k+R_j)\cdot X_s}, \tag{43}$$

where j_l is the l-th order spherical Bessel function. This implies that the coefficient $A_{lm}(k + R_j)$ is equal to :

$$A_{lm}(k + R_j) = 4\pi e^{i(k+R_j)\cdot X_s} i^l Y_{lm}\left(\frac{k+R_j}{|k+R_j|}\right) \frac{j_l(|k+R_j||R_{MT}^s|)}{u_l(E, R_{MT}^s)}. \tag{44}$$

The radius R_{MT}^s denotes here the radius of the considered MT sphere. This entirely determines our basis functions χ_j^k. Note that χ_j^k is continuous, but not derivable through the MT surface. This must be taken into account in the expression of the energy (hence in that of the hamiltonian) by adding some derivative jump terms [65].

Of course, the above technics are not the only ones. Let us mention the KKR (Korringa-Kohn-Rostoker) method [54, 52], which in some sense is an extension of the muffin-tin approximation, and uses the Green function method. The recursion method also uses Green function methods, but aims at calculating directly the DOS rather than solving the equations satisfied by the wave functions of the electrons. See for instance [40].

5.2   Reciprocal space integration and special point technics

As pointed out in Section 3.2, one frequently needs, in the course of electronic calculations, to compute the average value of some periodic function of the wave

---

[6]Hence the position of a nucleus, and in general the nearest one

vector k over the Brillouin zone (as for instance in (12)). In the present Section, we explain how the use of special point technics greatly reduces the computational cost of these evaluations.

For example, let f be a function defined on the reciprocal space, sharing the symmetries of the crystal. We want to evaluate

$$F = \int_{BZ} f(k)dk. \tag{45}$$

The first point is to expand f as a Fourier series, on the real space :

$$f(k) = \sum_{T \in \mathcal{R}} \hat{f}(T) e^{ikT}. \tag{46}$$

Next, note that if R is an element of the symmetry group of the crystal, then it is one of the reciprocal lattice, so that $f(Rk) = f(k)$, which implies that $\hat{f}(RT) = \hat{f}(T)$, since R is an orthogonal transformation. Hence, we may reorganize the above sum as follows :

$$f(k) = \sum_{n \in \mathbf{N}} f_n \frac{1}{\sqrt{\sharp C_n}} \sum_{T \in C_n} e^{ikT}, \tag{47}$$

where $(C_n)_{n \in \mathbf{N}}$ is a partition of the lattice vectors into non-symmetrically equivalent sets, n indexing them increasingly with respect to their norms. f_n is equal to $\sqrt{\sharp C_n} \hat{f}(T)$, for some $T \in C_n$. Hence, setting $A_n(k) = \frac{1}{\sqrt{\sharp C_n}} \sum_{T \in C_n} e^{ikT}$, which is normalized over the BZ, we have :

$$f(k) = \sum_{n \in \mathbf{N}} f_n A_n(k). \tag{48}$$

Since f_0 is exactly the evarage of f over the BZ, finding a k such that

$$\forall n \in \mathbf{N}, \ A_n(k) = 0, \tag{49}$$

would yield :

$$F = \int_{BZ} f(k)dk = f_0 = f(k). \tag{50}$$

Hence the evaluation of f would only require one evaluation of the function f, at a point which is *independent of f*. Of course, such a point does not necessarily exist in general. In particular, the number of integers for which (49) may be satisfied is limited by the compatibility between these equations. One thus needs to generalize this simple observation.

In a more general point of view, we look for a set of points $(k_i)_{1 \le i \le N_p}$ and a set of weights $(w_i)_{1 \le i \le N_p}$ such that

$$\sum_{i=1}^{N_p} w_i = 1, \quad \text{and} \quad \sum_{i=1}^{N_p} w_i A_n(k_i) = 0, \quad n = 1, \dots N_T. \tag{51}$$

Using the Fourier expansion of f, we then have :

$$\sum_{i=1}^{N_p} w_i f(k_i) = f_0 + \sum_{n=N_T+1}^{\infty} f_n \sum_{i=1}^{N_p} w_i A_n(k_i). \tag{52}$$

Thus, provided the last term is small enough, a good approximation to f_0 will be $\sum_{i=1}^{N_p} w_i f(k_i)$, the error being

$$\epsilon = - \sum_{n=N_T+1}^{\infty} f_n \sum_{i=1}^{N_p} w_i A_n(k_i). \tag{53}$$

In particular, the importance of this error is determined by that of the Fourier coefficients f_n, which decay very fast in the case of a smooth function. This makes this technic particularly powerful especially in the case of insulators. For metals, where the function f presents discontinuities (because the conduction band is only partially filled), one needs to use smearing technics, which basically consists in approximating f by a smooth function which integral is very close to that of f. See for example [37] or [16].

Hence, the main question is, how can one generate a set of special points? In many cases (see for instance [63, 18, 33, 7]) special points are only the smallest subset of a particular sampling of the BZ which image by symmetries of the crystal reproduce the corresponding sampling.

5.3   Ewald sums

As pointed out in Section 4, an important difficulty consists in giving a sense to the Coulomb potential (and the exchange term in HF theory). Since the second one is in some sense easier to deal with than the first one [28], we only consider here the first one.

In order to define the Coulomb potential V of a periodic distribution of charge $h = \sum_{j \in \mathcal{R}} h_0(\cdot + j)$, the first natural way is to set:

$$V(x) = \sum_{j \in \mathcal{R}} \left(h_0 \star \frac{1}{|x|} \right)(x + j). \tag{54}$$

The problem is that, if $\int h_0 = \int_{\Gamma(\mathcal{R})} h \neq 0$, the above sum, of which each term behaves like $\dfrac{\int_{\Gamma(\mathcal{R})} h}{|j|}$, is divergent. Thus, the only way to obtain a convergent series is to group the terms so as to have a neutral distribution of charge on each periodic cell. In the case of a neutral crystal, of which we are interested in, this corresponds to compute the potential of the electrons and the nuclei together rather than one at a time. Even in this case, the definition is not that clear[7]. Thus, we need a more efficient definition of V. It is provided by the Poisson equation:

$$-\Delta V = 4\pi h. \tag{55}$$

---

[7]In the case of a cubic unit cell, with h_0 shearing its symmetries, the quadrupole moment cancels, and the sum defined in (54) is convergent. In the general case, it is not.

This equation, with f defined as above, admits a unique $\mathcal{R}$-periodic solution up to a constant (as soon as $\int h_0 = 0$). The point is, such a V may be written as:

$$V = G \star_{\Gamma(\mathcal{R})} h, \tag{56}$$

where G is the periodic solution of

$$\begin{cases} -\Delta G = 4\pi \left(\sum_{j \in \mathcal{R}} \delta_j - \frac{1}{|\Gamma(\mathcal{R})|} \right), \\ G \quad \text{is} \quad \mathcal{R}-\text{periodic}, \end{cases} \tag{57}$$

and $\star_{\Gamma(\mathcal{R})}$ denotes the convolution product over $\Gamma(\mathcal{R})$, that is:

$$\left(h \star_{\Gamma(\mathcal{R})} g \right)(x) = \int_{\Gamma(\mathcal{R})} h(y)g(x - y)dy. \tag{58}$$

The potential G, which is uniquely defined up to a constant by (57), is thus the Coulomb potential associated to the lattice $\mathcal{R}$.

In other words, a good knowledge of G would allow to compute any potential arising from a periodic neutral distribution of charge. One straightforward way to compute this potential is to use the Fourier transform. Indeed, using the first equation of (57), one easily proves that:

$$|k|^2 \widehat{G}(k) = 4\pi, \quad \forall k \in \mathcal{R}^* \setminus \{0\}. \tag{59}$$

Here, the Fourier transform is defined by:

$$\widehat{f}(k) = \int_{\Gamma(\mathcal{R})} f(x)e^{-ik \cdot x}dx, \tag{60}$$

for all $k \in \mathcal{R}^*$. Thus, we have the following expression for G:

$$G(x) = C_0 + \frac{4\pi}{|\Gamma(\mathcal{R})|} \sum_{k \in \mathcal{R}^* \setminus \{0\}} \frac{e^{ik \cdot x}}{|k|^2}. \tag{61}$$

Here C_0 is an arbitrary constant, accounting for the fact that (57) defines G up to a constant. A common choice for C_0 is 0, i.e $\int_{\Gamma(\mathcal{R})} G = 0$, but it is not the only possibility. The point is that the above sum is only conditionally convergent, and thus, although useful for theoretical purposes, **not convenient** for computations. Hence, a more suitable expression of G must be obtained. It is obtained through the Ewald potential [34, 26]. It consists in screening G with a Gaussian, i.e in writing:

$$G(x) = G \star_{\Gamma(\mathcal{R})} g + (G - G \star_{\Gamma(\mathcal{R})} g), \tag{62}$$

where g is a periodized Gaussian:

$$g(x) = \frac{4\gamma^{3/2}}{\sqrt{\pi}} \sum_{j \in \mathcal{R}} e^{-\gamma |x-j|^2}, \quad \gamma > 0. \tag{63}$$

The trick is to develop the first term of (62) in Fourier series, because the Fourier transform of a Gaussian is a Gaussian ($\hat{g}(k) = e^{-\frac{|k|^2}{4\gamma}}$), and to develop the second one in direct space:

$$G \star_{\Gamma(\mathcal{R})} g(x) = \frac{1}{|\Gamma(\mathcal{R})|} \sum_{k \in \mathcal{R}^*} \hat{g}(k)\hat{G}(k)e^{ik \cdot x} = \frac{1}{|\Gamma(\mathcal{R})|} \sum_{k \in \mathcal{R}^*} \frac{4\pi}{|k|^2} e^{-\frac{|k|^2}{4\gamma} + ik \cdot x}, \tag{64}$$

$$(G - G \star_{\Gamma(\mathcal{R})} g)(x) = -\frac{\pi}{\gamma|\Gamma(\mathcal{R})|} + \sum_{j \in \mathcal{R}} \frac{\mathrm{erfc}(\sqrt{\gamma}|x - j|)}{|x - j|}. \tag{65}$$

with

$$\mathrm{erfc}(t) = 1 - \frac{2}{\sqrt{\pi}} \int_0^t e^{-u^2} du = \frac{2}{\sqrt{\pi}} \int_t^\infty e^{-u^2} du, \tag{66}$$

which behaves like $\frac{e^{-t^2}}{\sqrt{\pi}t}$ as $t \to \infty$. Collecting the above equation, we may thus write G as a sum over the direct lattice $\mathcal{R}$ whose terms behave like $\frac{e^{-\gamma|j|^2}}{\sqrt{\pi}|j|^2}$, plus a sum over $\mathcal{R}^*$ whose terms behave like $\frac{e^{-\frac{|k|^2}{4\gamma}}}{|k|^2}$:

$$G(x) = -\frac{\pi}{\gamma|\Gamma(\mathcal{R})|} + \sum_{j \in \mathcal{R}} \frac{\mathrm{erfc}(\sqrt{\gamma}|x - j|)}{|x - j|} + \frac{4\pi}{|\Gamma(\mathcal{R})|} \sum_{k \in \mathcal{R}^* \setminus \{0\}} \frac{e^{-\frac{|k|^2}{4\gamma} + ik \cdot x}}{|k|^2}. \tag{67}$$

(The first term is present to satisfy the normalization $\int_{\Gamma(\mathcal{R})} G = 0$.) The parameter γ is to be set so as to have a good compromise between the convergence of the two sums: the first sum converges better when γ is large, whereas the second one converges better when γ is small. Furthermore, this expression of G allows to assert:

Proposition 5.1 Let $h(x) = \sum_{j \in \mathcal{R}} h_0(x + j)$ be a periodic distribution of charge, such that:

(a) g_0 is fast-decaying at infinity (i.e, $|x|^p h_0 \in L^1(\mathbf{R}^3)$, $\forall p \geq 1$);

(b) the multipole moments of order $0, 1, 2$ of f_0 cancel.

Then, defining $F = G \star h_0 = G \star_{\Gamma(\mathcal{R})} h$, F is the Coulombic potential generated by h, i.e satisfies (57). Moreover, F satisfies:

$$F(x) = \sum_{j \in \mathcal{R}} \left(\frac{\mathrm{erfc}(\sqrt{\gamma}|x - j|)}{|x - j|} \star h_0 \right) + \frac{4\pi}{|\Gamma(\mathcal{R})|} \sum_{k \in \mathcal{R}^* \setminus \{0\}} \left(\frac{e^{-\frac{|k|^2}{4\gamma} + ik \cdot x}}{|k|^2} \star h_0 \right), \tag{68}$$

and these two sums are normally convergent.

This gives a way to numerically compute the Coulombic potential of such a distribution, by truncating the above sums, provided one can find such an h_0. A proof of this (provided h is neutral) may be found in [44]. Detailed discussions on these topics may be found in [36], appendix B, or in [26].

Note that the condition (a) is only technical, whereas condition (b) is more important: if (b) is not satisfied, equation (68) computes the Coulombic potential of the projection of h on the subspace of functions satisfying (b).

Of course, this basic formula must be then used in an efficient way, by grouping together terms so that the convergence is increased, and choosing the truncation of the series in a suitable manner. We refer to [28], chapter 2, for an example of these kind of considerations.

In the case of many atoms per cell, the above method should be improved by the use of the Reduced-Cell Multipole Method (RCMM), which consists, roughly speaking, in grouping the particles together into bigger and bigger sets as one goes away from the point at which we compute the potential. See [23, 24] for a description of this method, which is an adaptation to the periodic case of the standard fast-multipole method [41].

6   Conclusion

We have displayed here the roots of numerical solid state quantum chemistry, insisting on periodic-type codes [25, 38, 64, 75]. These methods have experienced lots of advances in the recent years, becoming nowadays in position to reach the same accuracy and speed as the more widely used cluster approachs, which roughly speaking consist in isolating a cluster of matter and treating it through molecular technics, thereby benefitting from these methods, which are much older and more accurate.

The near future will surely give birth to new and more powerful technics, as for example N-scaling methods (see [71], chapter 3 or [57]), or RCMM method [23].

Bibliography

[1] Andersen O.K., *Linear methods in band theory*, Phys. Rev. B 12, p 3060, 1975.

[2] Ashcroft N.W., Mermin N.D., *Solid state physics*, Holt-Saunders International Editions, 1976.

[3] Aprà E., Dovesi R., Freyria Fava C., Harrison N.M., Roetti C., Saunders V.R., *Ab-initio Hartree-Fock treatment of ionic and semi-ionic compounds: state of the art*, Phil. Trans. R. Soc. Lond. A 341, p 203, 1992.

[4] Aprà E., Dovesi R., Prencipe M., Saunders V.R., Zupan A., *Ab-initio study of the structural properties of LiF, NaF, KF, LiCl, NaCl, and KCl*, Phys. Rev. B 51, p 3391, 1995.

[5] Arbman G.O., Koelling D.D., *Use of energy derivative of the radial solution in an augmented plane wave method: application to copper*, J. Phys F 5, p 2041, 1975.

[6] Bachelet G.B., Hamann D.R., Schlüter M., *Pseudopotentials that work: from H to Pu*, Phys. Rev. B 26, p 4199, 1982.

[7] Baldereschi A., *Mean-value point in the Brillouin zone*, Phys. Rev. B 7, p 5212, 1973.

[8] Balian R., *From microphysics to macrophysics; methods and applications of statistical physics*, II, Springer, 1991.

[9] Baroni S., Pastori Parravicini G., Pezzica G., *Quasiparticle band structure of lithium hybride*, Phys. Rev. B 32, p 4077, 1985.

[10] Becke A.D., *Density-functional exchange energy approximation with correct asymptotic behavior*, Phys. Rev. A 38, p 3098, 1988.

[11] Biagini M., *Self-interaction corrected density-functional formalism*, Phys. Rev. B 49, p 2156, 1994.

[12] Blaha P., Schwarz K., *Electron densities and chemical bonding in TiC, TiN, and TiO derived from energi band calculations*, Int. J. Quant. Chem. 23, p 1535, 1983.

[13] Bloch F., *Über die Quantenmechanik der Elektronen in Kristallgittern*, Z. Phys. 52, p 555, 1928.

[14] Board J.A.Jr, Humphres C.W., Lambert C.G., Rankin W.T., Toukmaji A.y., *Ewald and multipole methods for periodic N-body problems*, in Lecture Notes in Computational Science and Engineering, 4, *Computational molecular dynamics: challenges, methods, ideas*, Deuflhard P., Hermans J., Leimkuhler B., Mark A.E., Reich S., Skeel R.D. (Eds), Springer, 1999.

[15] Board J.A.Jr, Toukmaji A., *Ewald sum techniques in perspective: a survey* Comput. Phys. Comm., 95, p 73, 1996.

[16] Boon H.M., Methfessel M.S., Mueller F.M., *Singular integrals over the Brillouin zone: the analytic-quadratic method for the density of states*, J. Phys. C 19, p 5337, 1986.

[17] Boys S.F., *Electronic wave functions I. A general method of calculation for the stationary states of any molecular system*, Proc. Roy. Soc. A 200, p 542, 1950.

[18] Chadi D.J., Cohen M.L., *Special points in the Brillouin zone*, Phys. Rev. B 8, p 5747, 1973.

[19] Cancès E., *SCF algorithms for HF electronic calculations*, in this volume.

[20] Catto I., Le Bris C., Lions P.-L., *Recent mathematical results on the quantum modelling of crystals*, in this volume.

[21] Chevary J.A., Fiolhais C., Jackson K.A., Pederson M.R., Perdew J.P., Singh D.J., Vosko S.H., *Atoms, molecules, solids, and surfaces: applications of the generalized gradient approximation for exchange and correlation*, Phys. Rev. B 46, p 6671, 1992.

[22] Conolly J.W.D., *The Xα method*, Semi-empirical methods of electronic structure calculation, Segal G.A. (Ed.), Plenum Press, 1977.

[23] Ding H.-Q., Karasawa N., Goddard W.A., *Atomic level simulations on a million particles: the cell multipole method for Coulomb and London nonbond interactions*, J. Chem. Phys. 97, p 4309, 1992.

[24] Ding H.-Q., Karasawa N., Goddard W.A., *The reduced cell multipole method for Coulomb interactions systems with million-atom unit cells*, Chem. Phys. Lett. 196, p 6, 1992.

[25] Dovesi R., Causà M., Harrison N.M., Orlando R., Pisani C., Roetti C., Saunders V.R., Zicovich-Wilson C.M., *CRYSTAL 98*, 1998. Available at http://www.ch.unito.it/ifm/teorica/crystal.html.

[26] Dovesi R., Freyria-Fava C., Roetti C., Salasco L., Saunders V.R., *On the electrostatic potential in crystalline systems where charge density is expended in Gaussian functions*, Mol. Phys. 77, p 629, 1992.

[27] Dovesi R., Pisani C., Roetti C., Saunders V.R., *Treatment of Coulomb interactions in Hartree-Fock calculations of periodic systems*, Phys. Rev. B 10, p 5781, 1983.

[28] Dovesi R., Pisani C., Roetti C., *Hartree-Fock ab-initio treatment of crystalline systems*, Lectures Notes in Chemistry 48, Springer, 1988.

[29] Eastham M.S.P., *The Spectral Theory of Periodic Differential Equations*, Scottish Acad. Press, Edinburgh-London, 1973.

[30] Eastham M.S.P., *The Schrödinger equation with a periodic potential*, J. Proc. R. Soc. Edinburgh, Sect. A 69, p 125, 1971.

[31] Ebbsjö I., Kalia R.K., Li W., Nakano A., Vashishta P., *Molecular dynamics methods and larg-scale simulations of amorphous materials*, in Amorphous Insulators and Semiconductors, p 151, edited by Thorpe M.F., Mitkova M.I., Kluwer academic publishers, 1997.

[32] P. Engel, *Geometric crystallography. An axiomatic introduction to crystallography*, R. Reidel Publishing Company, 1942.

[33] Evarestov R.A., Smirnov V.P., *Special points of the Brillouin Zone and their use in the solid state theory*, Phys. Stat. Sol. 119, p 9, 1983.

[34] Ewald P.P., *Die Berechnung optischer und elektrostatischer Gitterpotentiale*, Ann. Phys. (Leipzig) 64, p 253, 1921.

[35] Freeman A.J., Krakauer H., Weinert M., Wimmer E., *Full-potential self-consistent linearized-augmented-plane-wave method for calculating the electronic structure of molecules and surfaces: O_2 molecule*, Phys. Rev. B 24, p 864, 1981.

[36] Frenkel D., Smit B., *Understanding Molecular Simulation. From Algoritms to Applications*, Academic Press, 1996.

[37] Fu C.L., Ho K.M., *First principles calculation of the equilibrium ground-state properties of transition metals: applications to Nb and Mo*, Phys. Rev. B 28, p 5480, 1983.

[38] Furthmüller J., Hafner J., Kresse G., *VASP*, 1999. Available at http://cms.mpi.univie.ac.at/vasp/.

[39] Fulde P., *Electron Correlations in Molecules and Solids*, Solid-State Sciences, Vol 100, Springer, Berlin, 1993.

[40] Gibson A., Haydock R., LaFemina J.P., *Ab initio electronic structure computations with the recursion method*, Phys. Rev. B 47, p 9229, 1993.

[41] Greengard L., Rokhlin V., *A fast algorithm for particle simulations*, J. Comp. Phys. 73, p 325, 1987.

[42] Gunnarson O., Jones R.O., *Density functionnal calculations for atoms, molecules and clusters*, Rev. Mod. Phys. <u>61</u>, p 689, 1989.

[43] Hamann D.R., *Generalized norm-conserving pseudopotentials*, Phys. Rev. B <u>40</u>, p 2980, 1989.

[44] Harris F.E., *Theoretical Chemistry: Advances and Perspectives* <u>1</u>, edited by Eyring H. and Henderson D., 1975.

[45] Hay P.J., Wadt W.R., *Ab initio effective core-potentials for molecular calculations. Potentials for the transition metal atoms Si to Hg*, J. Chem. Phys. <u>82</u>, p 270, 1985.

[46] Hay P.J., Wadt W.R., *Ab initio effective core-potentials for molecular calculations. Potentials for main group elements Na to Bi*, J. Chem. Phys. <u>82</u>, pp 284, 1985.

[47] Hay P.J., Wadt W.R., *Ab initio effective core-potentials for molecular calculations. Potentials for K to Au including the outermost core orbitals*, J. Chem. Phys. <u>82</u>, pp 299, 1985.

[48] Hippert F., Gratias D., *Lectures on quasicrystals*, Les Éditions de la Physique Les Ulis, 1994.

[49] Hohenberg P., Kohn W., *Inhomogeneous electron gas*, Phys. Rev. B <u>136</u>, p 864, 1964.

[50] Johnson B.G., Gonzales C.A., Gill P.M.W., Pople J.A., *A density-functional study of the simplest hydrogen abstraction reaction. Effect of self-interaction correction*, Chem. Phys. Lett. <u>221</u>, p 100, 1994.

[51] Kittel C., *Introduction to Solid State Physics*, 6th Edition, Wiley and Sons, 1986.

[52] Kohn W., Rostoker N., *Solution of the Schrödinger equation in periodic lattices with an application to metallic Lithium*, Phys. Rev. <u>94</u>, p 1111, 1954.

[53] Kohn W., Sham L.J., *Self-consistent equations including exchange and correlation effects*, Phys. Rev. A <u>140</u>, p 1133 (1964).

[54] Korringa J., *On the calculation of the energy of a Bloch wave in a metal*, Physica <u>13</u>, p 392, 1947.

[55] Kuchment P., *Floquet Theory for Partial Differential Equations*, Operator Theory: Advances and Applications. 60. Basel: Birkhaeuser Verlag, 1993.

[56] Lee C., Parr R.G., Yang W.Y., *Developement of the Colle-Salvetti correlation energy formula into a functional of the electron density*, Phys. Rev. B <u>37</u>, p 785, 1988.

[57] Lee T.S., Yang W., *A density-matrix divide-and-conquer approach for electronic structure calculations of large molecules*, J. Chem. Phys. 103, p 5674, 1995.

[58] de Leeuw S.W., Perram J.W., Smith E.R., *Simulation of electrostatic systems in periodic boundary conditions I. Lattice sums and dielectric constants*, Proc. R. Soc. London A 373, p 27, 1980.

[59] de Leeuw S.W., Perram J.W., Smith E.R., *Simulation of electrostatic systems in periodic boundary conditions II. Equivalence of boundary conditions*, Proc. R. Soc. London A 373, p 56, 1980.

[60] de Leeuw S.W., Perram J.W., Smith E.R., *Simulation of electrostatic systems in periodic boundary conditions III. Further theory and applications*, Proc. R. Soc. London A 388, p 177, 1983.

[61] Loucks T.L., *Augmented Plane Wave Method*, Benjamin, New York, 1967.

[62] Methfessel M., Paxton A.T., *High-precision sampling for Brillouin zone intergration in metals*, Phys. Rev. B 40, p 3616, 1989.

[63] Monkhorst H.J., Pack J.D., *Special points for Brillouin-zone integration*, Phys. Rev. B 7, p 5212, 1976.

[64] MSI San Diego, *CASTEP*, 1999.

[65] Nemoshkalenko V.V., Antonov V.N., *Computational Methods in Solid State Physics*, Gordon Beach Publishers, 1998.

[66] Ostlund N.S., Szabó A., *Modern Quantum Chemistry*, Mc Graw-Hill, New York (1993).

[67] Parr R.G., Yang W., *Density-Functional Theory of Atoms and Molecules*, Oxford University Press, 1989.

[68] Perdew J.P., *Density-functional approximation for the correlation energy of the inhomogeneous electron gas*, Phys. Rev. B 33, p 8822, 1986 (errata: ibid 34, p 7406, 1986).

[69] Perdew J.P., Wang Y., *Accurate and simple analytic representation of the gas correlation energy*, Phys. Rev. B 45, p 13244, 1992.

[70] Perdew J.P., Zunger A., *Self-interaction correction to density-functional approximations for many-electrons systems*, Phys. Rev. B 23, p 5048, 1981.

[71] Pisani C. (Ed.), *Quantum-mechanical Ab-initio Calculation of the Properties of Crystalline Materials*, Lecture Notes in Chemistry 67, Springer, 1996.

[72] Popov V.A., *Solution of the Hartree-Fock equations for electrons in a crystal*, Comp. Math. & Math. Phys., 38, No. 6, p 937, 1998.

[73] Reed M., Simon B., *Methods of Modern Mathematical Physics*, IV, Academic Press, 1978.

[74] Senechal M., *Quasicrystals and geometry*, Cambridge University Press, 1996.

[75] Te Velde G., *ADF 2.3*, Vrije Universiteit, Amsterdam, 1997. Available at http://www.scm.com/ or http://www.csc.fi/chem/progs/adf/adf.html.

[76] Vanderbilt D., *Soft self-consistent pseudopotentials in a general eigenvalue problem*, Phys. Rev. B 41, p 7892, 1990.

Chapter 8

Examples of hidden numerical tricks in a solid state determination of electronic structure.

Mireille Defranceschi [1,2] & **Vanina Louis-Achille** [1,3]
[1] Commissariat à l'Energie Atomique, IPSN/DPRE/SERGD,
BP 6, F-92265 Fontenay-aux-Roses Cedex, France
[2] *CE-Saclay, Bâtiment 125,*
DPE/SPCP/LEPCA,
91191 Gif-sur-Yvette Cedex, France.
defrancesc@carnac.cea.fr
[3] *CNRS-CECM, F-94147 Vitry-Sur-Seine, France.*
Louis-Achille.Vanina@glvt-cnrs.fr

Abstract: The paper focuses on the numerical aspects of the atomistic modelling of materials in the Density Functional formalism. As usual in such a modelling, calculations can be run at various levels of theory and using different numerical options. After a review of the approximations underlying the Local Density Approximation (LDA), the Local Spin Density Approximation (LSDA), etc. the numerical options regarding the size of the basis functions, the use of pseudo-potentials or the selection of k-points are examined in details as well as the subsequent numerical tricks occuring in the calculations, e.g. in geometry optimizations. Finally, results regarding an apatitic mineral are reported and orders of magnitude of the expected accuracy are provided.

1   Introduction

Atomistic computer modelling methods are now well established tools in contemporary science [1]. They have been used for long in the study of molecules and are now used routinely, in conjunction with related experiments, in the field of materials chemistry even for industrial purposes. The scope of materials science is to monitor their macroscopic properties. It is now recognise that the richness of materials chemistry finds its fundamental explanation in the interactions of the nuclei and the electrons of the atoms constituting the solid. Therefore the electronic structure plays a key role in the understanding and prediction of a large number of phenomena. Quantum mechanics has put a mathematical framework to these interactions several decades ago but the explosion of numerical techniques coupled with the development of computers allow nowadays quantitative predictions of materials properties. Starting with quantum mechanics which describes atomistic phenomena, it is also possible to describe mesoscopic physics by using classical continuum mechanics and on top of all this statistical mechanics links atomistic processes with macroscopic properties.

However even if such methods have been used extensively in a considerable amount of situations ranging from metals to glasses, a lot of points remain questionable from both mathematical and numerical points of view. In the present paper we shall present an example of determination of the electronic structure of a mineral : the fluoroapatite $Ca_{10}(PO_4)_6F_2$ for which we shall focus on some numerical points necessary to deal with in standard quantum chemistry calculations. The first part of the paper highlights the fundamental concepts of the density functional theory which is the theory used in the present calculations. The second part is dedicated to a discussion of specific computational methods such as the pseudo-potential approach, plane wave methods, etc, it constitutes the central part of this article.

2   General aspects of the Density Functional Theory (DFT)

The time-independent Schrödinger equation :

$$\hat{H} |\Psi\rangle = E |\Psi\rangle,\qquad(1)$$

where $\hat{H}$ is the Hamiltonian operator describing a particular chemical system containing A nuclei, of charge Z_A and located at positions R_A, N electrons, of elementary charge e is solved within the so-called Born-Oppenheimer approximation (which assumes that the electrons adjust instantly to any change of the nuclei due to their much smaller mass than that of the nuclei). Ψ is the many electron wave function :

$$\Psi = \Psi(1, 2, ..., N)\qquad(2)$$

where $(1, 2, ..., N)$ denotes the cartesian coordinates and the spin coordinates of each electron. In the DFT the total energy of a system is expressed as a functional of the electronic density, ρ, in its ground state [3]:

$$E = E[\rho]. \tag{3}$$

A detailed review of functionals is given in [4]. Moreover ρ can be expressed [2] from wave functions of effective non-interacting electrons, Ψ_i . such that they generate the exact density of the interacting many electron system :

$$\rho = \sum_i n_i \Psi_i^* \Psi_i \tag{4}$$

The DFT formulation of the Schrödinger equation is therefore :

$$E[\rho] = T[\rho] + U[\rho] + \int V_{eff}(r)\rho(r)dr \tag{5}$$

where the kinetic energy, T, writes :

$$T[\rho] = \sum_i n_i \int \Psi(r) \left(-\frac{h^2}{2m}\nabla^2\right) \Psi^*(r)dr \tag{6}$$

and $U[\rho]$, the Coulomb energy is the sum of the electrostatic interactions between the electrons and the nuclei and the repulsion between all electrons and all nuclei :

$$U[\rho] = U_{en} + U_{ee} + U_{nn} \tag{7}$$

$$U[\rho] = -e^2 \sum_\alpha^A \int Z_\alpha \frac{\rho(r)}{|r - R_\alpha|} dr + e^2 \int \int \frac{\rho(r)\rho(r')}{|r - r'|} dr dr' + e^2 \sum_{\alpha\alpha'} \frac{Z_\alpha Z_{\alpha'}}{|R_\alpha - R_{\alpha'}|} \tag{8}$$

The last term of (7) corresponds to the exchange and correlation energy, E_{xc}. Contrary to the other two terms of (5) which can be easily determined, the E_{xc} term must be approximated.

The most simple approximation for the exchange and correlation energy E_{xc} consists in considering that it depends only on the local electron density around each volume element, it is consequently called the Local Density Approximation (LDA) :

$$E_{xc}[\rho] = \int \rho(r)\epsilon_{xc}^{hom}[\rho(r)] dr \tag{9}$$

it assumes that the exchange and correlation effects come mainly from the close neighbours and that the density is relatively homogeneous.

In some circumstances where the electronic density is inhomogeneous it is necessary to go beyond the LDA, considering the gradient of the density. In this approximation, the exchange and correlation energy E_{xc}, instead of being written as in (9), writes as :

$$E_{xc}[\rho] = \int \rho(r)f_{xc}[\rho(r), \nabla\rho(r)] dr \tag{10}$$

For systems where the electronic density changes strongly, various analytical forms exist for $E_{xc}[\rho(r)]$. Readers interested in various analytical expressions of E_{xc} can refer to [1].

Finally when a good approximation to E_{xc} is found one can apply the variational principal to (5) assuming (4) is valid. This leads to :

$$\frac{\partial E}{\partial \Psi_i} = 0 \qquad (11)$$

or :

$$\left[-\frac{1}{2}\nabla^2 + V_e + \mu_{xc} \right] \Psi_i = \varepsilon_i \Psi_i \qquad (12)$$

which are the Kohn Sham equations, where V_e is the Coulomb potential and μ_{xc} the exchange and correlation operator.

So far, equations have been given ignoring spin functions. However for some practical applications, the number of electrons with spin-up can be different from that with spin-down and a local spin density (LSD) has been developed where the exchange and correlation potential becomes dependent of the spin. The spin polarized Kohn Sham equations can be written as [5] :

$$\{ -\frac{h^2}{2m}\nabla^2 + V_e + \mu_{xc}\left[\rho(r), \sigma(r)\right] \} \Psi_i^\sigma(r) = \varepsilon_i^\sigma \Psi_i^\sigma(r) \; with \, \sigma = \uparrow or \downarrow \qquad (13)$$

where the total electronic density, $\rho(r)$, is :

$$\rho(r) = \rho_\uparrow(r) + \rho_\downarrow(r) \qquad (14)$$

and the spin density, $\sigma(r)$, is :

$$\sigma(r) = \rho_\uparrow(r) - \rho_\downarrow(r) \qquad (15)$$

Depending on the system under consideration, the formalism might be even more complicated : relativistic correction might be included, and the potential terms can be periodic or non-periodic, etc.

At this point of the description, it is worth stressing that DFT methods which are usually considered as ab initio methods because no experimental parameters are needed in the calculations are not strictly speaking ab initio since the exchange and correlation terms are approximated.

3   Choice of the method

Since the central task of electronic structure calculations is to solve the effective one-particle equations deriving from the Schrödinger equation, running DFT calculations necessitates not only to select the needed level of theory (LDA, LSD, GGA, etc) but also to choose between various numerical options. All the possibilities will not be discussed here since it would be too long but some very precise examples will be considered.

3.1   Representation of the function

As usual, the representation of the wave functions in terms of basis functions is central in quantum chemical calculations since it conditions the accuracy and consequently the cost of the calculation. The basis can be an atomic one and the atomic orbitals can be either numerical or constructed from gaussian or Slater functions . For periodic systems, wave functions can be written as a product of a periodic function, $\xi_i(r)$, and a phase term, e^{ikr}, where k is a Bloch vector (or wave vector) belonging to the first Brillouin zone (the Brillouin zone is defined as a small polyhedron centered on the origin of the reciprocal space limited by the bissectrices of the directing vectors, G_1, G_2, G_3. The Bloch vector verifies the so-called Bragg condition :

$$|(k + G)|^2 = |k|^2 \tag{16}$$

reducing the number of wave functions from a continuous to a discrete set of basis functions. The periodic term, $\xi_i(r)$ can be expanded in terms of plane waves :

$$\xi_i(r) = \sum_G C_{i,G}^k e^{iGr} \tag{17}$$

where G is a vector of the reciprocal space :

$$G = n_1 G_1 + n_2 G_2 + n_3 G_3 \tag{18}$$

The sum in (17) is expanded on the whole reciprocal space and is infinite, then each wave function writes :

$$\Psi_i^k(r) = \sum C_{i,k+G} e^{i(k+G)r} \tag{19}$$

Consequently the electronic wave functions are developed on a discrete plane wave basis for each k-point of the Brillouin zone. The expansion is doubly infinite : infinite number of plane waves and infinite number of k-points. Practically infinite calculations are not runable. Two physical observations allow to circumvent this difficulty ; i) for close values of k electronic wave functions do not differ very much, the wave function for a given zone of the k space can be described by the wave function for an appropriate value of k, ii) coefficients $C_{i,k+G}$ of the plane waves with a low kinetic energy (given by $\frac{1}{2}|(k + G)|^2$) are significantly higher than those corresponding to plane waves with a high kinetic energy. The plane wave basis can be truncated keeping only those with a kinetic energy smaller than a selected cut-off energy,E_c [6]:

$$\frac{1}{2}|(k + G)|^2 < E_c \tag{20}$$

The electronic wave functions are therefore determined for a finite number of k-points of the Brillouin zone and expanded in terms of a finite number of plane waves. Depending on the oscillating character of the function the number of plane waves needed might be very important, for instance, a f function needs a much greater number of plane waves than a s function. The trade is then accuracy (i.e. number of plane waves and cut-off energy) versus computation time and memory allocation.

In Fig. 8.1 are given the total energies of two solids (YPO_4 in grey and $ScPO_4$ in black) ; greater is the cut-off energy more precise are the results. Practically for these systems to be sure to obtain a converged value the chosen cut-off must be at least equal to 60 Ha (1630 eV). It is worth pointing out that the convergence of energy differences occurs at a much smaller cut-off energy than that required for absolute convergence. This is easily understood if one remembers that the plane waves of highest values of k change the electronic wave functions only within the ion cores and not in the bonding regions. However, running energy calculations with a reduced cut-off leads to a ragged result in the energy curve, and here again a good compromise is to be looked for.

3.2   Use of pseudo-potentials

One inconvenience of using plane waves is that they are not suited for oscillating functions. For the wave functions describing core electrons which present numerous oscillations, the use of pseudo-potentials allows to avoid the difficulty. Inside spheres centered on the various nuclei the highly oscillating potential is replaced by a smooth pseudo-potential while out of the spheres the correct potential is considered. The pseudo-potential has to satisfy some obvious mathematical constraints (e.g. continuity of the functions and of their two first derivatives onto the spheres) but also some physical constraints (the pseudo-atom and the real atom must have the same behavior regarding the chemical bonds and for some other cases their chemical hardnesses have to be identical). Physically speaking using a pseudo-potential is equivalent to have a pseudo-atom built from a positive ion and part of its electrons (considered as core electrons) which are frozen and the rest of the electrons (considered as valence electrons). Accordingly the pseudo-atom are described by pseudo-wave functions (for the various levels s, p, d, etc) and the system is described by a pseudo-energy which has no physical meaning but differences between pseudo-energies can be compared with physical quantities. The first task for generating pseudo-potentials is then to select for each type of atom the radii of the spheres for the s, p, d, etc levels which have to include all the oscillations of the potential but leaving out of the spheres the last extremum. Then after selecting a parametrized expression for the pseudo-potential, the various constraints are applied yielding to different categories of pseudo-potentials. In usual calculations, the user of pseudo-potentials only has to select, from physical considerations, which type of pseudo-potential his problem needs (e.g. a simple Norm Conserving (NC) potential or an Extended Norm and Hardness Conserving (ENHC) one or even an Ultra-Soft one), the radii of the spheres are therefore fixed and the user has nothing else to do numerically. In Fig. 8.2, are represented pseudo-wave functions and a Norm Conserving pseudo-potential for the silicon atom (N=14) where only four electrons are explicitly treated in the following electronic configuration $3s^2 3p^3 3d^0$ with the radii accidentally equal to 2.09 a.u. for all the orbitals.

4   Choice of the number of k-points

The calculations described above involve integrating periodic functions of a Bloch wave vector over the Brillouin zone. To minimize the calculations, it is convenient to compute the functions for a selected set of k-points as small as possible. Methods have been developed for the selection of special points which partly rely on symmetry considerations. In the present calculations, the method developed by Monkhorst and Pack [7] has been employed. From a practical point of view, the only thing to do for the user is to determine how many points are necessary to obtain accurate results. The general trend is that for metals many points are needed while for insulating compounds a few k-points yield realistic values. As a matter of fact for $ScPO_4$ and YPO_4, all other parameters being identical, varying from 3 to 6 k-points changes the total energy by less than 0.01%.

5   Other numerical tricks

One of the drawback of the choice of a finite plane wave basis set is that the number of basis functions changes discontinuously with the cut-off energy. Consequently, running a geometry optimization a discontinuity might happen in the curve giving the total energy of the system as a function of the cell volume (Fig.8.3). The stress tensor (which is related to the derivative of the energy with the geometry) might present spurious oscillations too. For each increase (or decrease) of the cell volume, the G values in the reciprocal space decrease (or increase) accordingly for a given cut-off, the number of plane waves defined by k, changes discontinuously. A new set (for each G vector) of (k+G) vectors might be added (or removed) to the basis set, the total energy discontinuously decreases (or increases) due to the added (removed) variational freedom in the wave function. It is always possible to reduce the magnitude of the error by increasing the value of the cut-off energy or by using a denser k-point set. However, the problem remains, it can be handled by applying a correction factor which accounts approximately for the difference between the number of states in a basis set with infinitely large number of k-points and the number of basis states actually used in the calculation [8].

The correction to the energy is :

$$E_{corrected} = E_{calculated} - \frac{\partial E_{calculated}}{\partial \ln(N^{pw})} \cdot \frac{N^{pw}}{V_{cell} \cdot g_\eta} \tag{21}$$

and :

$$g_\eta = \frac{2^{3/2}}{6\pi^2} \cdot E_c^{3/2} \tag{22}$$

where N^{pw} is the geometric average of plane waves for each k point calculated by the code, and V_{cell} is the unit cell volume. The product $V_{cell} \cdot g_\eta$ corresponds to an ideal number of plane waves which takes no account of the discrete nature of the number of plane waves. To calculate the energy derivative, three calculations are performed at a fixed geometry including the lattice constants close to the expected value of the final geometry. Then two other calculations are run for a cut-off energy increased or decreased by 3% (changing the number of plane waves which are considered). Then the total energy versus $\ln(N^{pw})$ is calculated. Finally the corrected value of the energy is obtain by (21.

A possible way out is to work with a fixed number of plane waves ; however, it is generally observed that choosing a constant cut-off energy leads to smaller errors in static equilibrium [9].

Similarly calculations regarding the stress tensor are affected by the change of number of plane waves. In the stress tensor, close to the equilibrium geometry, all the non-diagonal components are close to zero, the diagonal components are written as :

$$\sigma_{ii} = \frac{cell(i)}{V} \cdot \frac{dE_{calc.}}{dcell(i)} \tag{23}$$

where $i = x, y, z, cell(i)$ is an array containing the cell parameters and V is the volume of the cell. The difference between the calculated stress corresponding to a locally constant N^{pw} and the corrected stress corresponding to a strictly constant E_{cut} [9] is defined as the Pulay stress. Thus :

$$\sigma_{ii} = \frac{cell(i)}{V_m} \cdot \frac{\partial E_{calc.}}{\partial cell(i)} + \frac{1}{V_m} \cdot \frac{\partial E_{calc.}}{\partial N^{op}} \tag{24}$$

The last term is obtained as above from three calculations of total energy with different N_{pw} at a fixed volume. The inclusion of this correction yields physical properties much closer to the experimental values than is otherwise possible at this cut-off energy.

5.1   Results

After having selected the level of theory needed for a problem and after having chosen the various numerical parameters entering in the numerical procedure, a real calculation can be run for fluoroapatite, $Ca_{10}(PO_4)_6F_2$, which crystallizes in the hexagonal structure according to the space group $P6_{3/m}$ with a=b=9.398 Å and c= 6.878 Å , there is one $Ca_{10}(PO_4)_6F_2$ formula (42 atoms) per unit cell.

The calculations are based on LDA-DFT calculations with the valence orbitals expanded in plane wave basis sets. Exchange and correlation are included through the local density approximation, using the Ceperley and Adler potential as parameterized by Teter [10]. The full ionic potential of the solid is replaced with a much softer pseudo-potential generated using the procedure proposed by Troullier and Martins [11]. The minimization of the Kohn-Sham energy functional is realized with a conjugate-gradients technique [6]. A single k-point and an energy cut-off equal to 1400 eV are taken. Energy and stress correction factors are applied throughout the calculations.

Prior to any calculations in quantum chemistry one has to perform a geometry optimization of the system under study. Two parameters are sufficient to determine an hexagonal structure : the two lattice parameters a and c, or equivalently the cell volume V and the ration c/a. In the sequel, the total energy is first minimized with

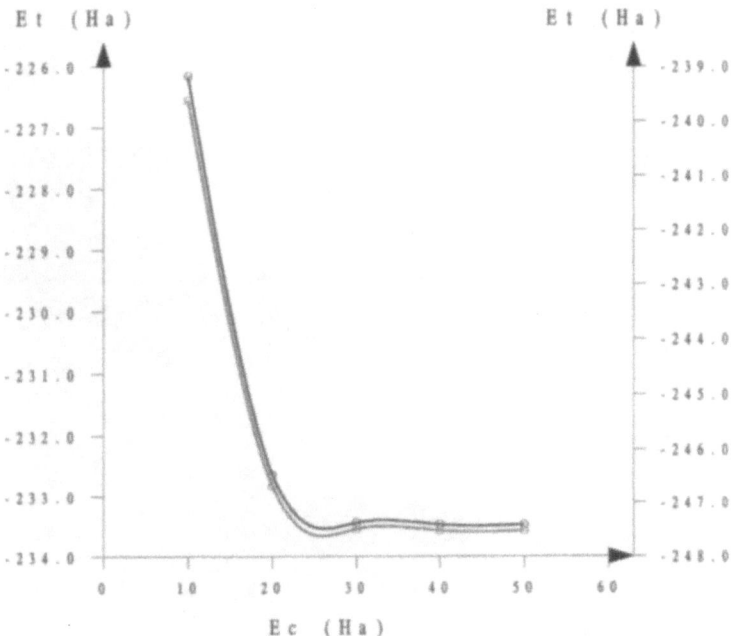

Figure 8.1: Total energy computed with respect to cut-off energy E_c for YPO_4 (grey) and $ScPO_4$ (black)

respect to the cell volume V and the atomic coordinates are relaxed using Hellmann-Feyman forces as a guide. The c/a ratio is calculated under the condition that the stress tensor is isotropic. The computed values are respectively a=b=9.375 Å and c=6.844 Å. They are within 0.3% of the experimental structures. The calculated structures are slightly more contracted than the experimental one, underestimation of lattice parameters is inherent to the LDA approximation.

Modeling of mineral matrices as complex as apatites are very seldom found in the literature. It is quite a challenge, with nowadays computers, to reconcile good quality results and CPU. The present study, although being done on a very simple apatite, shows that monitoring correctly the various numerical parameters, allows to perform calculations which can be useful for the determination of structural parameters and related physical quantities of inorganic solids. Works in progress [12] on a lanthanide doped fluor-britholite structure (a silicated apatite) show that the procedure developed in the present paper can be extrapolated to other more sophisticated solids.

Acknowledgements. The authors wish to express their thanks to L. De Windt (CEA-IPSN) for his stimulating comments and ideas.

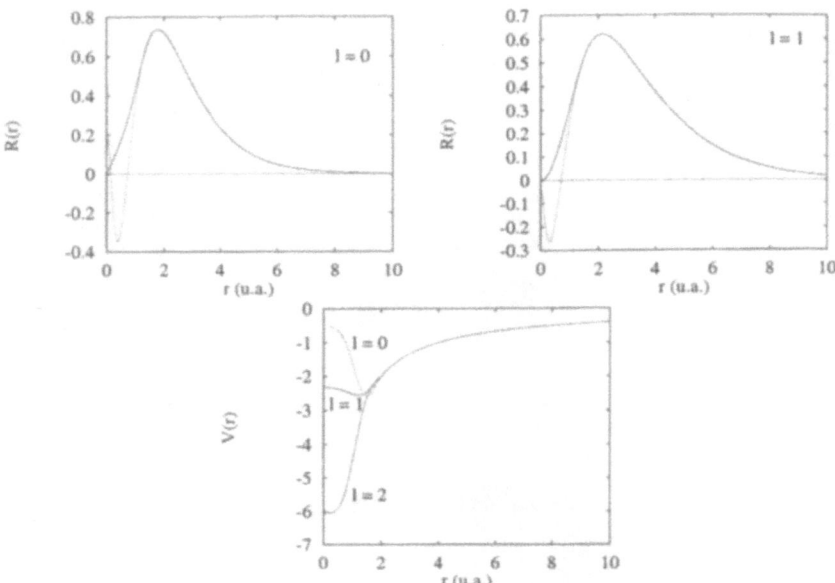

Figure 8.2: Construction of the pseudo-wave functions, $R(r)$, (pseudo-wave functions are black, and real wave functions in grey) and of the pseudo–potential, $V(r) = V_{ion}(r)$, for Si. $r_c = 2.208\ u.a.$ for all l value.

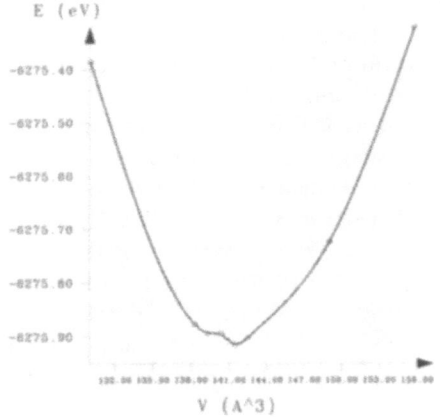

Figure 8.3: Total pseudo–potential energy of YPO_4 with respect to volume.

Bibliography

[1] C. R. A. Catlow, G. Bell, J. D. Gale. *Accounts. Chem. Res.*, 12 (1979) 276.

[2] W. Kohn, L. J. Sham. *Phys. Rev. A*, 140 (1965) 1133.

[3] P. Hohenberg, W. Kohn. *Phys. Rev. B*, 136 (1964) 864.

[4] R. O. Jones, O. Gunnarsson. *rev. Mod. Phys.*, 61 (1989) 689 .

[5] S. H. Vosko, L. Wilk, M. Nusair. *Can. J. Phys.*, 58 (1980) 1200 .

[6] M. C. Payne, M. P. Teter, D. C. Allan, T. A. Arias, J. D. Joannopopoulos. *Rev. Mod. Phys.* 64 (1992) 1045.

[7] H. J. Monkhost, J. D. Pack. *Phys. Rev. B*, 13 (1976) 5188.

[8] G. P. Francis, M. C. Payne. *J. Phys. : Condens. Matter*, 2 (1990) 4395 .

[9] P. Dacosta, O. H. Nielsen, K. Kunc. *J. Phys. C*, 19 (1986) 3163.

[10] M. P. Teter. *unpublished but available in the Plane_Wave 3.0.0 code*, MSI, San Diego, 1995

[11] N. Troullier, J. L. Martins. *Phys. Rev. B*, 43 (1991) 1993.

[12] L. Boyer, V. Louis-Achille, J. Carpena, M. Defranceschi, L. De Windt, J.L. Lacout, *Amer. Min.* (to be published).

Chapter 9

Quantum mechanical models for systems in solution

Benedetta Mennucci
Dipartimento di Chimica e Chimica Industriale,
Università di Pisa, via Risorgimento 35,
56126 Pisa, Italy
bene@dcci.unipi.it

Abstract : An overview of modern theories for the modelling of solvent effects on the state and the properties of quantum mechanical molecular systems is presented. The emphasis here is on the models that exploit a continuum description of the solvent and introduce effective Hamiltonians to represent intermolecular interactions. The main theoretical and numerical aspects of these methods, in which mutual interactions between solute and solvent are included, are presented and discussed.

As more specialistic feature, we also analyze their extension to the derivative theory, presenting some selected applications such as the search for the best geometry and the evaluation of molecular response properties in solution. In this context some comments on the eventual inclusion of dynamical aspects are also reported.

1   Introduction

The importance of getting a good understanding of surrounding media effects on chemical systems is difficult to overestimate. Applications go from condensed phase chemistry, biochemical reactions in vitro to biological systems in vivo.

With the enormous progress achieved by computing technology, an increasing number of models and phenomenological approaches are being used to describe the effects of a given surrounding medium on the properties of a selected subsystem. A number of quantum chemical methods and computational codes, usually applied to calculate *in vacuo* systems, have been supplemented with a variety of solvation model representations [1].

With the increasing number of methodologies applied to this important field, it is becoming more and more difficult for non-specialist to cope with theoretical developments and applications. The present exposition may thus represent a useful starting point for anyone who wants to enter in this research field. However, due to the huge number of related aspects, a preliminary choice on the range of subjects which shall be covered has to be done.

In general, theoretical studies on the effects of the solvent on the properties and behavior of molecules are performed according to a large variety of methods; however, for our scopes, it is useful to classify them according to two basically different strategies. In the first, which we shall not treat at all, we can collect supermolecule calculations and computer simulations. Even if the philosophy of the two approaches is quite different, in both cases, a detailed description of the disposition and structure of molecules composing the liquid system is searched.

The second strategy, almost complementary to the previous one, collects methods in which a target subsystem, the "solute" (eventually supplemented by a few nearby solvent molecules) is described at the microscopic level, and a secondary subsystem ("the solvent") is, on the contrary, modeled as an infinite macroscopic continuum medium having suitable properties.

There are several approaches to be classified in this second domain. Some among them belong to the category of semiempirical methods and some play a role in the study of complex solutes (e.g. molecules of biological interest). Others retain the formalism of the *in vacuo ab initio* molecular calculations, including in the Hamiltonian, and in the corresponding Schrödinger equation, an explicit expression of the solute-solvent potential. The latter approach, also known with the name Effective Hamiltonian Method, is that we shall almost exclusively focus on in the present contribution.

2   The Effective Hamiltonian and the Free Energy

In the Effective Hamiltonian method the solvent S is represented by a homogeneous continuum medium which is polarized by the solute M placed in a cavity built in the bulk of the dielectric. The solute-solvent interactions are described in terms of a solvent reaction potential. The basic hypothesis of this kind of models is that one can always define a free energy functional, $\mathcal{G}(\Psi)$,[2, 3, 4, 5] depending on the solute electronic wavefunction Ψ. This energy functional can be expressed in the following general form:

$$\mathcal{G}(\Psi) = \langle \Psi | \hat{H}^0 | \Psi \rangle + \langle \Psi | \hat{\rho}_r | \Psi \rangle V_r^R + \frac{1}{2} \langle \Psi | \hat{\rho}_r | \Psi \rangle \mathbf{V}_{rr'}^R \langle \Psi | \hat{\rho}_{r'} | \Psi \rangle \tag{1}$$

In (1) the Born-Oppenheimer (BO) approximation is employed. This means a standard partition of the Hamiltonian into an electronic and a nuclear part, as well as the factorization of the wave function into an electronic and a nuclear component. In this approximation (1) refers to the electronic wavefunction with the electronic Hamiltonian dependent on the coordinates of the electrons, and, parametrically, on the coordinates of the nuclei.

The detailed description of the various terms comparing in the equation above will be given in the following, here it suffices to say that $\hat{H}^0$ is the Hamiltonian describing the isolated molecule, $\hat{\rho}_r$ represents the operator of the solute electronic charge density, V_r^R is the *solvent permanent potential*, and $\mathbf{V}_{rr'}^R$ describes the *response function of the reaction potential* associated with the solvent. Here an extension of the Einstein convention on the sum has been exploited: the space variables r and r', appearing as repeated subscripts, imply an integration in the 3-dimensional space.

Until the system is considered at the absolute zero temperature, $\mathcal{G}$ is the internal energy of the solute-solvent system. On the contrary, if the quantities V_r^R and $\mathbf{V}_{rr'}^R$ are defined for a model at non-zero temperature, $\mathcal{G}$ has to be considered as a free energy.

By applying the variational principle on this functional we can derive the nonlinear Schrödinger equation specific for the system under scrutiny. If we impose that first-order variations of $\mathcal{G}$ with respect to arbitrary variations of the solute wavefunction Ψ are zero, and that the latter is normalized, the nonlinear Schrödinger equation becomes:

$$\hat{H}_{eff} | \Psi \rangle = \left[\hat{H}^0 + \hat{\rho}_r V_r^R + \hat{\rho}_r \mathbf{V}_{rr'}^R \langle \Psi | \hat{\rho}_{r'} | \Psi \rangle \right] | \Psi \rangle = E | \Psi \rangle \tag{2}$$

where E is the Lagrange multiplier introduced to fulfill the normalization condition on the electronic wavefunction. The equation (2) defines the specific 'Effective Hamiltonian', $\hat{H}_{eff}$, giving the name to the whole procedure.

The first solvent term, V_r^R is often assimilated to an average solvent potential, it can also play the role of an instantaneous solvent potential with respect to a dynamical behavior. The presence of such perturbative term in the molecular Hamiltonian does not lead to any difficulty, neither from the theoretical point of view, neither from the practical. Many examples are known in which an external potential is introduced in the molecular calculations. On the contrary, the treatment of the reaction potential operator $\hat{\rho}_r \mathbf{V}_{rr'}^R \langle \Psi | \hat{\rho}_{r'} | \Psi \rangle$ is rather delicate, as this term induces a nonlinear character to the solute Schrödinger equation.

As last preliminary note we recall that the nonlinear equation (2) is a direct consequence of the variational principle applied to $\mathcal{G}$. The (free) energy functional $\mathcal{G}$ has a privileged role in the theory, as the solution of the Schrödinger equation gives a minimum of this functional even though it is not the eigenvalue of the nonlinear Hamiltonian, here indicated as E. We stress that in the habitual linear Hamiltonians these two quantities, the Hamiltonian eigenvalue and the variational functional, coincide. The difference between E and $\mathcal{G}$ has, however, a clear physical meaning; it represents the polarization work that the solute does to create the charge density inside the solvent. It is worth remarking that this interpretation is equally valid for zero-temperature models and for those in which the thermal agitation is implicitly or explicitly taken into account.

To pass from the free energy functional defined above to the thermodynamical analog, further items have to be considered. First we have to choose the reference state (in our case it is given by non-interacting nuclei and electrons of M, supplemented by the unperturbed, i.e. unpolarized, pure liquid S), and to take into account all the contributions not present in the reaction potential; by collecting the latter in a single term, indicated as G_{nR} the total free energy can be written as:

$$
\begin{aligned}
G &= \langle \Psi | \hat{H}^0 | \Psi \rangle + \langle \Psi | \hat{\rho}_r | \Psi \rangle V_r^R + \frac{1}{2} \langle \Psi | \hat{\rho}_r | \Psi \rangle \mathbf{V}_{rr'}^R \langle \Psi | \hat{\rho}_{r'} | \Psi \rangle \\
&\quad + \frac{1}{2} E_{NN} + V_{NN} + G_{nR}
\end{aligned}
\tag{3}
$$

where Ψ is now the electronic wavefunction obtained as solution of the Schrödinger equation (2) and V_{NN} the nuclear repulsion energy of the solute M. The term E_{NN}, here introduced with factor $1/2$ in order to have a free energy, represents the interaction energy between solute nuclei and the solvent reaction component generated by the nuclear part of the solute charge distribution.

The derivation above shows that solvent effects can be divided into two main classes; the first, to be defined in a self consistent way with solute charge distribution, acts both on Ψ and on the free energy functional, and the other, to be evaluated independently from the QM calculation, contributes only to the evaluation of G.

Remark that the definition of G given in (3) does not contain contributions due to thermal motions of the molecules. On the contrary, in order to get a complete definition of the free energy of solvation, we should supplement the given definition of G with entropic contributions, so that [6]:

$$
G^S = G - RT \ln \left(q_{rot,s} q_{vib,s} \right) + RT \ln \left(n_{M,s} \Lambda_{M,s}^3 \right)
\tag{4}
$$

Here $q_{rot,s}$ and $q_{vib,s}$ are the microscopic partition functions for rotation and vibration of M in the solution; while $n_{M,s}$ and $\Lambda_{M,s}$ indicate the related numeral densities and the momentum partition functions, respectively.

In the computational practice, use is often made of a 'reduced' solvation energy, $G - E^0$, in which the difference between the molecular motion contributions calculated *in vacuo* and in solution is neglected. The error done in this approximation is almost due to the magnitude of the change in the rotational contribution passing from the gas-phase to the solution; in fact, by choosing as the two standard states, two infinitely large systems with the same density and temperature, the term of translation nature vanishes and the change in the vibrational contribution is generally limited.

3   Intermolecular Forces

The computational strategies formulated to get the complete free energy are numerous and of different level of accuracy. In general, use is made of a modellization of the solvent interactions derived from the theory of intermolecular forces. In this framework the reaction response scheme introduced above, and the following energetic quantity G, is written as a sum of contributions of different physical origin, related to dispersion, steric and electrostatic forces among solute and solvent molecules. The formulation of an explicit reaction response function for each specific interaction is a very difficult task if the model has to contemporaneously fulfill the request of being complete, physically stated, and computationally feasible. What can be achieved with a smaller effort is an "hybrid" description in which the various terms are treated at different levels of accuracy and within independent interpretative models. This can be obtained by classifying the interactions in terms of their relative importance with respect to the characteristics of the given system. In this way the theoretical and computational effort will be focussed on the main contributions, leaving the description and the evaluation of the others to simpler procedures. Following this strategy, the nonlinear QM system introduced in (2) will be limited to treat only the more effective part of the whole solute-solvent interaction, while the rest can be computed independently on the QM calculation, by resorting to separate calculations based on empirical parameters or other simplifying assumptions.

Generally speaking, the classification of the various interactions, namely of dispersive, steric and electrostatic origin, is given in terms of polarity and polarizability considerations on both solute and solvent molecules. Thus, electrostatic forces will be taken as dominant in highly polar solvent and polar or ionic solutes, while the always present dispersion term can be assumed to become really important in the presence of molecules whit no free charges or electrical dipole moments. This simple analysis becomes a little more complex for the term indicated as steric contribution.

The steric contribution is conveniently divided into two terms: one defined only by the shape and the dimension of the solute considered as a rigid body, and the other related to the Pauli exclusion principle. While the latter can be easily introduced in the previous scheme in terms of intermolecular forces as the so-called 'repulsion contribution', the former has to be treated in a different framework. This contribution, more often called 'cavitation', corresponds to the work spent in building up a cavity of appropriate shape and volume in which the solute molecule is enclosed, with all the other solute-solvent interactions switched off. Being related to classical forces among rigid bodies, its effects are only on the free energy of the system and not on the solute wave function, and its always positive contribution to G can be evaluated independently on the QM calculation.

For the evaluation of G_{cav} several formulas are available, based on the shape and size of the solute and on different parameters of the solvent: surface tension [7], surface tension with microscopic corrections [8], isothermal compressibility [9], and geometrical data of the molecules [10]. The first three formulas here mentioned are empirical and follow almost the same philosophy of the continuum dielectric, neglecting the discrete nature of the solvent molecules but making use of experimental bulk parameters. The last formulation, on the contrary, derives from a theory based on a discrete model of fluids (the Scaled Particle Theory, SPT) [10]; however, also in this case the final expression of G_{cav} depends on bulk solvent parameters only.

3.1   Repulsion and Dispersion

What we call here repulsion contribution is the second term of the steric interaction introduced in the previous subsection. It is here collected together with the dispersion contribution, although of different physical origin, as their analysis and the consequent classification can be given in a unified way.

In general, the modelling of dispersion and repulsion interactions in solution is based either on a discrete molecular description of the liquid or on a continuum model.

The discrete approach is generally based on the use of pair potentials related to atoms or groups of atoms of the solvent S (here indicated with l) and the solute M (here indicated with m):

$$V_{dis-rep} = \sum_{m \in M} \sum_{l \in S} V_{ml}(r_{ml}) \iff V_{ml}(r_{ml}) = \sum_n \frac{d_{ml}^{(n)}}{r_{ml}^{(n)}} \tag{5}$$

the dispersion (n=6,8,10) and the repulsion (n=12) coefficients are taken from the literature. Often an alternative exponential expression, more related to the physical interpretation of the interaction, is used for the repulsion term: $c_{ml} \exp(-\gamma_{ml} r_{ml})$.

As for the previous cavitation term this potential is completely independent on solute charge distribution; it is not involved in the QM description of the system, but it affects only the total free energy value.

The second possible approach we have mentioned to get $G_{dis-rep}$ is based on a continuum model; in this case the two contributions are treated separately.

One again, the starting point is the general expression derived from the intermolecular forces. In the previous discrete approach however, this equation were applied to calculated or estimated potentials V_{rep}^{AB} available from the literature. We now substitute V_{rep}^{AB} for a suitable expression taken directly from the theory. The original idea is to consider that, as originated from the Pauli exclusion principle, the repulsion forces between two interacting molecules increase with the overlap of the two distributions and are strictly related to the density of electrons with the same spin [12]:

$$G_{rep} = \frac{1}{2}\rho_B \int g_{AB}(r)dr \int \frac{dr_1 dr_2}{r_{12}} P_A(r_1; r_2) P_B(r|r_2; r_1) \tag{6}$$

Here the label A refers to the solute, B to the solvent, r is an appropriate set of coordinates defining the internal geometry of the complex AB, ρ_B is the number density and g_{AB} is a correlation function which is 0 inside the solute cavity ($r \in C$) and 1 outside ($r \notin C$). At this point, because in the continuum approach here exploited the electron density of the solvent is not given, it is useful to make the following two assumptions: (i) each valence electron pair of the solvent molecules can be localized in bond and lone pair regions and (ii) each pair, owing to the thermal motion of the solvent molecules, will have the same probability to be found at any point of the solution not occupied by the solute. The resulting solvent density $P_{pair}(R|r_2; r_1)$ (here R is the coordinate of the centroid of the localized orbital containing the pair in a reference frame fixed on the solute molecule) can be represented in terms of a Gaussian representation of localized orbitals; this yields the simple expression

$$G_{rep} = \alpha \int_{r \notin C} dr\, P_A(r) \tag{7}$$

where α is a suitable constant defined by some selected properties of the solvent.

We note that in (7) the repulsion energy is proportional to the fraction of solute electrons outside the cavity; as a consequence the minimization of the total free energy functional (1), when this contribution is included, provides an automatic confinement of the electronic cloud of the solute.

The continuum approach to the dispersion term has a much longer history than for the repulsion, and different procedures have been developed.

The original formal theory is expressed in terms of quantum electrodynamics and the continuum medium characterized by its spectrum of complex dielectrics frequencies [13]. A more recent formulation [14], derived from this theory, is based on the extension of the reaction field concept to a dipole subject to fluctuations exclusively electric in origin. In this framework, the dispersion free energy of a molecule placed in a cavity immersed in the solvent is related to the molecular polarizability and the complex dielectric constant. A couple of methods now in use [15, 16], have extended this treatment, in origin limited to the dipolar approximation and a spherical cavity, to a full multipolar expansion and to any cavity shape. Actually, the two methods are quite different either in the solvation model they exploit or in the practical implementation; anyway an important aspect is common to both: the inclusion of the dispersive term in the solute effective Hamiltonian which allows a real influence of these interaction forces on the electronic structure of the solvated molecule (i.e. on the description of its wave function).

Recently an alternative procedure [12] has been formulated starting, as for the repulsion contribution, from the theory of intermolecular forces. In this framework, the expression of the dispersion energy between two molecular systems A and B is given in terms of generalized frequency dependent polarizabilities. Following a scheme commonly exploited to derive the electrostatic contribution to the interaction energy, the molecule B is substituted by a continuum medium, the solvent, described by a surface charge density σ_B induced by the electric field of the solute A and spreading on the cavity surface.

Within this framework, the dispersion free energy becomes

$$G_{dis} = \frac{1}{\pi} \int_0^\infty d\omega \sum_{K \neq 0} \frac{\omega_{0K}^A}{(\omega_{0K}^A)^2 + \omega^2} \int dr_1 \times \tag{8}$$

$$\int_\Sigma \frac{dr_2}{r_{12}} P_A(0K|r_1) \sigma_B[\epsilon_B(i\omega), P_A(0K|r); r_2]$$

where $P_A(0K|r)$ and ω_{0K}^A are respectively transition densities and energies for the solute A (for transition to state K). As already said, σ_B is the surface charge density induced in the solvent by the electric field of the charge distribution $P_A(0K|r)$; it depends on a solvent dielectric constant calculated at imaginary frequencies, $\epsilon_B(i\omega)$. The reduction of the general expression (8) to an equation containing only terms related the solute ground state can be obtained through some simplifications on the form of σ_B and the nature of the excited states to be considered in the summation of (8).

3.2   Electrostatic contribution

In the previous sections devoted to the other interaction terms we have not entered too deeply inside the intrinsic nature of the dielectric, as the theory below each derivation did not require this kind of analysis. On the contrary, the electrostatic problem of a charge distribution embedded in a cavity surrounded by a continuum dielectric strongly depends on the macroscopic structural characteristics of the dielectric itself.

Within the continuum framework, this preliminary classification can be limited to the form of the dielectric constant. We can thus define the following systems: (i) homogeneous isotropic dielectrics, characterized by a constant scalar permittivity, ϵ; (ii) homogeneous anisotropic dielectrics, characterized by a constant tensorial permittivity, ϵ; (iii) inhomogeneous dielectrics, characterized by a position dependent permittivity, $\epsilon(r)$. Other specific dependencies can be found as well as complex systems formed by two or more dielectrics separated by well defined boundaries, but in general the three listed above cover almost all the most interesting cases. A different very interesting chemical problem not automatically connected to this list is that of ionic solutions; i.e. liquids in which charged particles are free to move.

This scheme cannot be exhaustive but a detailed analysis of all possible systems goes far beyond the scopes of the present work. What we shall do here is to focus our attention almost exclusively on the simplest, and also the most common case (a single solute in an infinitely dilute solution of a homogeneous isotropic solvent), referring to specific papers for all the details of the extension of the method to more complex systems.

For the selected system, i.e. a solute M in a cavity C surrounded by an infinite homogeneous isotropic dielectric with permittivity ϵ, the basic relation to be considered is the Poisson equation, with the related boundary conditions; namely:

$$\begin{cases} -\Delta V = 4\pi\rho_M & \text{in } C \\ -\epsilon\Delta V = 0 & \text{outside } C \\ [V] = 0 & \text{on } \Sigma \\ [\partial V] = 0 & \text{on } \Sigma \end{cases} \tag{9}$$

where V is the electrostatic potential due to the presence of the charge distribution ρ_M located inside the cavity and Σ the cavity surface. The jump condition $[V] = 0$, means that the potential V is continuous across the interface Σ, i.e. $V_e - V_i = 0$ on Σ, where the subscripts e and i indicate regions outside and inside the cavity, respectively. The equality $[\partial V] = 0$ is a formal expression of the jump condition of the gradient of the potential; in our limit isotropic system it takes the well-known form:

$$\left(\frac{\partial V}{\partial n}\right)_i - \epsilon \left(\frac{\partial V}{\partial n}\right)_e = 0 \tag{10}$$

where n is the outward pointing unit vector perpendicular to the cavity.

The main approaches used to solve this problem can be divided in different classes; in the following we shall focus our attention on the three most used:

1. the apparent surface charge (ASC) methods: PCM [17, 18], COSMO [19, 20], IEF [21];

2. the multipole expansion (MPE) methods: SCRF [22];

3. the generalized Born approximation (GBA) methods: AMSOL [23].

In the list above each class is associated with one or more specific computational procedures, indicated with their acronyms, which represent the most successful applications of the class itself. Let us start our analysis from the last one: the GBA methods, here represented by the AMSOL code.

GBA methods (AMSOL).

These methods can be defined as the generalization of the model Born formulated for the simple system of a point charge q placed at the center of a void spherical cavity with radius a (or equivalently a conducting sphere with net charge q and radius a). The extension of the Born formula of the electrostatic solvation free energy to a set of point charges placed in a general cavity can be written as follows:

$$G_P(Born) = -\frac{\epsilon - 1}{2\epsilon}\frac{q^2}{a} \Rightarrow G_P(GBA) = -\frac{\epsilon - 1}{2\epsilon}\sum_{k,k'} q_k q_{k'} \gamma_{kk'} \tag{11}$$

where k, k' label the atomic centers and $\gamma_{kk'}$ are Coulomb-type integrals.

The same formalism, with the required changes, has been adopted by Cramer and Truhlar [24] in a series of methods, now included in the AMSOL code, which belong to the same class of effective Hamiltonian exploiting continuum description of the solvent we have defined at the very beginning.

In these methods, called SMx where x denotes the parametrization of the specific solvation method, the partial atomic charges q_k to be used in the application of (11) are calculated either from a semiempirical wave function using a zero-overlap Mulliken population analysis or, more recently, from a class of charge models which map the Mulliken derived charges with others reproducing more accurately experimental gas-phase dipole moments.

In the most recent versions of SMx models, the solvation free energy is calculated as a sum of two terms:

$$\Delta G_{sol} = \Delta G_{ENP} + G_{CDS} \tag{12}$$

where the first term ΔG_{ENP} contains, on turn, two components: ΔE_{ENP}, which represents the change in the internal electronic kinetic and electronic-nuclear coulombic energies of the solute upon relaxation in solution, and G_P which is the free energy of the electric polarization as derived from the generalized Born approximation. The second term of (12), G_{CDS}, accounts both for the free energy of forming a cavity in the solvent to make room for the solute, and for the changes in dispersion interactions and solvent structure that always accompany the solvation process.

Let us forget the CDS term, whose form can be easily derived by generalizing the surface tension-derived cavitation formula we have quoted in the section dedicated to the steric interaction, and let us focus on the electrostatic term G_{ENP}.

The whole method is here based on a self-consistent reaction field formalism, in which the general QM equation to be defined is the following:

$$\mathbf{F} = \frac{\partial G}{\partial \mathbf{P}} \tag{13}$$

where $\mathbf{F}$ is the Fock matrix, G the free energy functional, and $\mathbf{P}$ the solute density matrix. For a restricted Hartree-Fock calculation in which core electrons are not treated explicitly, the free energy functional becomes:

$$G = \frac{1}{2} \sum_{\mu,\nu} P_{\mu\nu}(h_{\mu\nu} + F_{\mu\nu}^{(0)}) + \sum_{i<j} \frac{Z_i Z_j}{R_{ij}} - \frac{\epsilon - 1}{2\epsilon} \sum_{k,k'} q_k q_{k'} \gamma_{kk'} \tag{14}$$

where μ and ν run over the set of valence atomic orbital functions, and Z_i is the nuclear charge of atom i minus its number of core electrons. Here $\mathbf{P}$ is the relaxed (with respect to solvation) density matrix, $\mathbf{F}^{(0)}$ the Fock matrix as defined *in vacuo*, but formed using the relaxed density matrix, and the final term accounts for electric polarization using the generalized Born formalism.

To evaluate the relaxed density matrix, one must solve for the orbitals with a proper Fock matrix as defined by (13). We underline that by explicitly including the polarization effects in the Fock operator, the resulting density matrix and converged orbitals are determined self-consistently in the presence of the solvent.

AMSOL [23] methods are fast, applicable to different solvents, and provided with many options, among which geometry optimizations.

MPE methods (SCRF)

This class of methods will be described in terms of a specific model known as the SCRF method of Nancy group [22], being it without any doubt the most complete MPE method.

As the previous GBA class is defined as the generalization of the Born model, the present one can be seen as the parallel extension of the Kirkwood's model. Here, in fact, the charge distribution of the solute placed in a cavity surrounded by a continuum dielectric is expanded in a series of multipoles. In this framework, if with M_m^l we indicate the m component of the multipole of order l written in the spherical tensor formalism, the electrostatic contribution to the free energy variation becomes:

$$\Delta G_{el} = -\frac{1}{2} \sum_{l=0}^{\infty} \sum_{m=-l}^{l} R_l^m M_l^m \tag{15}$$

in which R_m^l is a component of the reaction field, i.e. the corresponding derivatives of order l of the electrostatic potential created by the continuum polarized under the influence of the solute.

Within the linear response approximation the various components of the reaction field can be written in the form:

$$R_l^m = \sum_{l',m'} f_{l\,l'}^{m\,m'} M_{l'}^{m'} \tag{16}$$

in which the coefficients $f_{l\,l'}^{m\,m'}$, called the reaction field factors, only depend on the shape of the cavity and of the dielectric constant of the solvent.

In the original Kirkwood treatment, which exploits a spherical cavity, these factors have a simple analytical definition: they are independent from m and are nonzero only when $l = l'$. The analytical treatment is possible also for a spheroid or an ellipsoid. In the general case, the reaction field factors have to be computed numerically.

The formalism given above for a one-center expansion can be, and actually it has been, straightforwardly generalized to a distributed multipole expansion of N centers; this development, as expected, leads to much more rapidly convergent calculations of electrostatic solvation free energies. However, for simplicity's sake, in the following we shall continue to limit the exposition to the one-center expansion formalism.

In order to pass to the quantum calculation, here limited to the description of the solute electronic wave function as a single antysimmetrized product of one-electron molecular spin-orbitals expanded over a finite set of atomic orbitals $\{\chi_\mu\}$, we have to minimize the free energy functional. The resulting modified Fock matrix has thus the $(\mu\nu)$ element of the form:

$$F_{\mu\nu} = F_{\mu\nu}^0 + \sum_{l,m} \sum_{l',m'} M_l^m f_{l\,l'}^{m\,m'} \langle\mu|\, S_{l'}^{m'}\, |\nu\rangle \tag{17}$$

in which $F_{\mu\nu}^0$ is the corresponding matrix element for the isolated molecule and $S_{l'}^{m'}$ a solid spherical harmonic which defines the corresponding multipole tensor element. The perturbation term of (17), because of its nonlinearity, depends on the density matrix $\mathbf{P}$, nevertheless, this dependence does not introduce any special difficulty since the same characteristic occurs in $F_{\mu\nu}^0$. The iterative scheme in the SCF computation is adapted to this situation.

This methodology, originally developed for the evaluation of the electrostatic free energy only and limited to the simple case of a spherical cavity, more recently has been generalized to cavities of arbitrary shape, and to more complex calculations. It is worth mentioning that in a recent version of the code it has been introduced the calculation of analytical derivatives of the free energy [25] and hence the possibility of geometry optimization procedures, and that the same formalism of the reaction field, factors allows a more comprehensive approach in which the dispersion term is added to the electrostatic term in the derivation of the Hartree-Fock equation [15]. The explicit definition of this dispersion contribution is:

$$\Delta G_{dis} = -\frac{1}{8}\frac{\Delta E_M \Delta E_S}{\Delta E_M + \Delta E_S}\sum_{l,m}\sum_{l',m'} f_{l\,l'}^{m\,m'}(\epsilon_\infty)A_{l\,l'}^{m\,m'} \qquad (18)$$

Here ΔE are transition energies (indices M and S stand for the solute and solvent molecules, respectively), and $f_{l\,l'}^{m\,m'}(\epsilon_\infty)$ the reaction field factors computed with the dielectric permittivity ϵ_∞, the square of the optical refractive index extrapolated to infinite wavelength. $A_{l\,l'}^{m\,m'}$ is a component of the solute multipole polarizability in the spherical tensor formalism which can be evaluated by a variational technique from the ground-state wave function of the molecule.

ASC methods (PCM, COSMO, IEF).

In all of the various methods chosen as representatives of this class, namely PCM, COSMO and IEF, the strategy is to solve the system (9) and get the related potential V by exploiting an integral equation formalism [26]. Actually, for one of the three methods, namely that known with the acronym COSMO, the model is a little different as the screening effects in the dielectric are replaced by the screening effects in a conductor. In other words, COSMO method is a solution of the Poisson equation designed for the case of very high ϵ, and it takes advantage of the analytic solution for the limit case of a conductor ($\epsilon = \infty$), for which the boundary condition reduces to $V = 0$ on the surface. Anyway, apart from these differences in the theoretical background, COSMO can be described exactly in the same way as the other ASC methods, and, in fact, in the following we shall present a single derivation for all the methods.

The solution of system (9) can be obtained by exploiting different mathematical approaches. In particular PCM and COSMO exploit a more standard approach based on Green functions, while IEF a second one, of recent implementation, using other operator functions derived from the theory of integral equations. The formal derivation of the two approaches is quite different, nevertheless it is possible, and useful for our scopes, to find a common basic description.

In both approaches the first step is the definition of the analytical expressions for the specific operators; their knowledge enables one to transform the first two equations in (9) into integral equations on the surface Σ, that can be easily solved with standard numerical methods. In this framework, the solution of system (9) is given by the sum of two electrostatic potentials, one produced by ρ_M and the other due to a surface charge distribution σ, placed on the interface, arising from the polarization of the dielectric medium:

$$V(x) = V_M(x) + V_\sigma(x) = \int_{R^3} \frac{\rho_M(y)}{|x-y|}dy + \int_\Sigma \frac{\sigma(s)}{|x-s|}ds \qquad (19)$$

where the integral on the first term is taken over the entire three-dimensional space. The surface charge has to fulfill the additional condition derivable from the Gauss theorem:

$$Q_\sigma(theo) = \int_\Sigma \sigma(s)ds = -\frac{\epsilon-1}{\epsilon}Q_M \qquad (20)$$

where Q_M is the total net charge of the solute M. The condition expressed by (20) is clearly a great advantage of the ASC formulation with respect to all the other methods we have described in the previous subsections, in fact, it gives an easy checking test on the quality of the model employed.

Once (19) and (20) have been thus defined, the only but non-trivial problem consists in the calculation of the screening, or apparent (from which the ASC acronym defining this class of methods), surface charge density $\sigma(s)$.

We recall that the numerical approach used to treat the ASC density is equal in all the methods; it exploits a partition of the surface into K small portions, called tesserae, of known area a_k, on which a constant charge density is assumed. In this framework which can be easily linked to the analogous techniques used in the fields of physics and engineering and known as the Boundary Element Method (BEM) [27], the integral form of $V_\sigma(x)$ in (19) is reduced to a finite sum running over the point charges representing the surface charge:

$$V_\sigma(x) = \sum_k^K \frac{q_k(s_k)}{|x - s_k|} \Longleftrightarrow q_k(s_k) = \sigma(s_k)a_k \qquad (21)$$

where vector s_k indicates the position on each tessera k where to evaluate the constant value of σ (usually, it identifies the center of the tessera and is called the representative point). A more detailed analysis of this partition will be given in the section we shall devote to the techniques developed to define the molecular cavity.

PCM In this method, developed in its first form in 1981 [17], but then almost completely redefined in 1995 [18], the apparent surface charge is expressed by the following classical electrostatic relation:

$$\sigma(s) = - \left(\frac{\epsilon - 1}{4\pi\epsilon} \right) E(s) \cdot \hat{n}(s) \qquad (22)$$

where E is the total electric field, i.e. the sum of the contributions, both computed on the inner side of the surface cavity, due to the solute charge distribution and to the ASC itself, respectively, and $\hat{n}$ is the outward normal unit vector to the cavity at position s.

Due to the dependence of $E(s)$ on $\sigma(s)$, (22) should be solved through an iteration procedure; however the same results can be obtained in the form of a set of linear equations. As said above, in the computational practice a discretization technique is used to represent the continuous $\sigma(s)$ into point charges; in this framework the basic equations defining the apparent point charges, below collected in the column matrix **q**, are reduced to the following:

$$\mathbf{D}\mathbf{A}^{-1}\mathbf{q} = -\mathbf{E}_n^M \qquad (23)$$

where $\mathbf{E}_n^M$ collects cavity-orthogonal components of the solute electric field. $\mathbf{D}$ is a square matrix, with size equal to the number K of surface tesserae, and elements determined by the value of the normal components of the field the apparent charges induce on each other (i.e. the solvent reaction field), and $\mathbf{A}$ is a square diagonal matrix, of same dimension as $\mathbf{D}$, with elements equal to the tesserae's areas: $A_{ii} = a_i$.

COSMO   As already mentioned, COSMO is based on a screening conductor theory [19, 20]. Screening in conductors can be handled much more easily than in dielectric media; in our framework of a solute M placed in a cavity inside the medium, now represented by a conductor, the boundary condition to be fulfilled is that the total potential V defined in (19) cancels out on the surface Σ of the cavity.

By following the strategy already used to get (23) in the PCM method, from (22) we can write the basic system giving the COSMO apparent charges in the following matrix formalism:

$$\mathbf{BA}^{-1}\mathbf{Q} = -\mathbf{V}^M \tag{24}$$

where the column matrix $\mathbf{V}^M$, of dimension equal to the number K of the tesserae, contains the solute electrostatic potential on each tessera, and the elements of the diagonal $(K \times K)$ matrix $\mathbf{B}$ are the value of the electrostatic potential the apparent charges induce on each other.

The capital letter used in (24) for the matrix containing the apparent charges indicates that the latter are here derived for a real conductor, for which $\epsilon = \infty$; if the COSMO model is used to simulate a solvent with dielectric constant ϵ, these apparent charges have to be scaled so that their sum obeys the Gauss law (see 20): a very effective way to do this is to multiply each charge by the factor $(\epsilon - 1)/\epsilon$.

As a final note we add that COSMO method is correctly defined, and indeed it well reproduces solute energies and properties, only for highly polar solvents like water, for which the difference between the two values of $(1\text{-}1/\epsilon)$ obtained with $\epsilon = 78.5$ and $\epsilon = \infty$ is only 1.3%. Nevertheless, in the computational practice COSMO is used also for the treatment of nonpolar solvents with $\epsilon \approx 2$, but in these cases it needs approximations which are physically by far less founded than those defined for the dielectric model, and not always easily acceptable.

IEF   The main specificity of the IEF formalism [21] is that, instead of starting from the boundary conditions as in the standard PCM, it defines the Laplace and Poisson equations describing the specific system under scrutiny, here including also anisotropic dielectrics, and ionic solutions, and it introduces the relevant specifications by proper mathematical operators [28]. The fundamental result is that, as anticipated in the introduction, the IEF formalism manages to treat structurally different systems within a common integral equation-like approach. In other words, the same considerations exploited in the isotropic PCM model leading to the definition of a surface charge density $\sigma(s)$ which completely describes the solvent reaction response, are still valid here, also for the two mentioned non-isotropic systems.

Passing directly to the computational implementation, we recall that as for the previous ASC procedures, also in the IEF we exploit a tessellation of the cavity surface into K tesserae, and approximate the charge density σ by a piecewise constant function (i.e. a function constant on each tessera). Also in this case, the relations giving the corresponding apparent charges q_k, can be expressed in the form of matrix equations:

$$\mathbf{CA}^{-1}\mathbf{q} = -\mathbf{g} \tag{25}$$

where $\mathbf{q}$ is the column vector containing the apparent charges on the single tesserae,

C is a $K \times K$ matrix , A the usual diagonal matrix of tesserae areas, and g a column vector depending on the solute electrostatic potential and the normal components of the corresponding field (both computed at the centers of the tesserae). Matrix C is composed by the combination of four distinct matrices, each corresponding to different operators specific of the solvent (isotropic, anisotropic, or constituted by an ionic solution). In passing we recall that in the limit of isotropic dielectrics (i.e. in the same framework in which PCM and COSMO apply) matrix g can be reduced to the potential matrix V^M, with consequent reduction of computational time.

As a final note, we recall, even if without reporting any formal derivation, that IEF represents the most general method among those reported above; its basic expression contains both the standard PCM (what we have to do in this case, is just some mathematical manipulations) and COSMO; the latter can be easily obtained from the isotropic version of IEF by assuming that $\epsilon \to \infty$. The detailed formal derivation of both statements can be found in refs. [21].

4   Quantum Mechanical Theory of Solvation: the ASC approach

As reported in section (2), the fundamental quantity to be considered in solvation models is the free energy functional defined in (1). By imposing that first-order variation of $\mathcal{G}$ with respect to an arbitrary variation of the solute wavefunction Ψ is zero, and following the standard scheme developed for the self consistent field theory in vacuo, we finally obtain the generalized Fock operator, $\hat{F}'$ and the corresponding equation from which the final wavefunction has to be derived; in a matrix formalism we have:

$$\mathbf{F}'\mathbf{\Psi} = \mathbf{\Psi}\varepsilon \;\Leftrightarrow\; F'\psi_i = \varepsilon_i \psi_i \tag{26}$$

where we have exploited the properties of 1-determinant wavefunctions to be invariant under unitary mixing of the spin-orbitals ψ_i and of the Hermitian matrix of Lagrangian multipliers ε to be brought into a diagonal form.

To go a step further we need to explicitly define the solvent terms. To do that, we prefer to shift to a more limited formalism by introducing the common finite-basis approximation and a closed shell system. In this approximation we can eliminate the spin dependence (the occupied spin-orbitals occur in pairs $\psi_K = \phi_K\alpha$, $\bar{\psi}_K = \phi_K\beta$ with a common orbital factor ϕ_K) and expand the molecular orbitals (MOs) as a linear combination of atomic orbitals (LCAO).

On performing the spin integration in the equations used so far, we easily find, for the free energy:

$$\mathcal{G} = \mathrm{tr}\mathbf{Ph}^0 + \frac{1}{2}\mathrm{tr}\mathbf{PG}^0(\mathbf{P}) + \mathrm{tr}\mathbf{Ph}^R + \frac{1}{2}\mathrm{tr}\mathbf{PX}^R(\mathbf{P}) \tag{27}$$

and, for the generalized Fock matrix:

$$\mathbf{F}' = \mathbf{h}^0 + \mathbf{G}^0(\mathbf{P}) + \mathbf{h}^R + \mathbf{X}^R(\mathbf{P}) \tag{28}$$

where $\mathbf{P}$ is the one-electron density matrix and $\mathbf{h}^0$ and $\mathbf{G}^0(\mathbf{P})$ are the elements used in standard calculations *in vacuo* which collect one- and two-electron integrals, respectively.

In (27-28) the solvent contributions, previously introduced as potential functions, have been translated into a form recalling the vacuum system, just to emphasize the parallelism between the two calculations; in fact, also solvent effects can be partitioned in one- and two-electron contributions, indicated as $\mathbf{h}^R$ and $\mathbf{X}^R(\mathbf{P})$, respectively (we prefer not to use $\mathbf{G}^R$ to indicate the solvent two-electrons term as it presents different characteristics with respect to the analog in vacuo, $\mathbf{G}^0(\mathbf{P})$).

The terms referring to repulsion and dispersion effects can be derived from the general procedures presented in the previous sections. More specifically, the analysis done for the repulsion term has shown the existence of a direct proportionality between the resulting effect, i.e. the free energy, and the portion of solute charge distribution outside the cavity q_{out}^A. By rewriting the repulsion free energy reported in (7) as the expectation value of the related operator we obtain that repulsion interactions contribute to the Fock matrix only through the one–electron matrix [12]:

$$\mathbf{h}^{rep} = \alpha \left[\mathbf{S} - \mathbf{S}^{(in)}\right] \tag{29}$$

where $\mathbf{S}$ is the overlap matrix and

$$\mathbf{S}_{rs}^{(in)} = -\frac{1}{4\pi} \int_\Sigma E_{rs}(r)dr \tag{30}$$

In the equations above E_{rs} indicate the integrals corresponding to the outward component of the electric field due to the distribution $\chi_r \chi_s$.

The dispersive contributions are more complex, and lead to both one– and two-electron interaction matrices, namely [12]

$$\mathbf{h}_{ru}^{dis} = -\frac{\beta}{2} \sum_{st} [rs|tu]\mathbf{S}^{-1}{}_{st} \tag{31}$$

$$\mathbf{X}_{ru}^{dis}(\mathbf{P}) = \frac{\beta}{2} \sum_{st} [rs|tu]\mathbf{P}_{st} \tag{32}$$

The integrals $[rs|tu]$

$$[rs|tu] = \frac{1}{2} \int_\Sigma dr \left[V_{rs}(r)E_{tu}(r) + V_{tu}(r)E_{rs}(r)\right] \tag{33}$$

are computed by exploiting the partitioning of the cavity surface in terms of the tesserae (V_{rs} and E_{rs} indicate the potential and the outward component of the field due to the elementary distribution $\chi_r \chi_s$) and β is a constant defined by the solvent.

In order to derive a parallel analysis for the last electrostatic term it is useful to partition the solute field (or potential) used to get the apparent charges (see 23-25) into its two sources, namely the solute nuclei and electrons. In this way we can derive a parallel partition of the apparent charges, $\mathbf{q}$, into two components defined as electrons- and nuclei-induced, respectively. In this framework also the solute-solvent interaction part of the free energy (3) may be divided into different contributions:

$$G_{M-S} = \frac{1}{2} \int \int \frac{\rho(r)\sigma(s)}{|r-s|} dr ds = \frac{1}{2} \left(E_{ee} + E_{eN} + E_{Ne} + E_{NN} \right) \tag{34}$$

where the first subscript refers to the component of the solute charge and the second to that of the ASC.

In the MO-LCAO approximation introduced above, the first three contributions to the interaction energy of (34) (the last term E_{NN} is a constant dependent only on the positions and the charges of solute nuclei) can be expressed in terms of matrices defined on the AOs.

Actually, the two mixed e-N interactions as formally equivalent can be cast in a single one-electron term, $\mathbf{h}^{el}$ [18]:

$$\mathbf{h}_{rs}^{el} = - \sum_k \mathbf{V}_{rs}(s_k) q^N(s_k) \tag{35}$$

while the e-e term can be associated to a pseudo two-electron matrix $\mathbf{X}$ defined as:

$$\mathbf{X}_{rs}^{el}(\mathbf{P}) = - \sum_k \mathbf{V}_{rs}(s_k) q^e(s_k; \mathbf{P}) \tag{36}$$

where we have explicitated the dependence of the electron-induced charges, q^e, on the density matrix $\mathbf{P}$.

By combining all the terms we obtain the generalized Fock equation:

$$\begin{aligned} \mathbf{F'T} &= \left[\mathbf{h}^0 + \mathbf{h}^{rep} + \mathbf{h}^{dis} + \mathbf{h}^{el} + \mathbf{G}^0(\mathbf{P}) + \mathbf{X}^{dis}(\mathbf{P}) + \mathbf{X}^{el}(\mathbf{P}) \right] \mathbf{T} \\ &= \left[\mathbf{F}^0 + \mathbf{F}^R \right] \mathbf{T} = \mathbf{ST}\varepsilon \end{aligned} \tag{37}$$

Equation (37) can be solved with the same iterative procedure of the problem *in vacuo*; the only difference introduced by the presence of the continuum dielectric is that, at each SCF cycle, one has to simultaneously solve the standard quantum mechanical problem and the additional problem of the evaluation of the interaction matrices. In this scheme the apparent charges are obtained through a self-consistent technique which has to be nested to that determining the solute wave function; as a consequence, at the convergency, solute and solvent distribution charges are mutually equilibrated.

5 The solute cavity and its tessellation

In all the various approaches described above, one defines an empty cavity in the dielectric medium in which the solute M resides. The shape and size of the cavity are critical factors in the elaboration of a method. A cavity with a wrong shape introduces distortions in the description of the reaction field and of the related solvent effects.

The cavity shapes actually employed can be classified as follows: (1) regular shapes (i.e. spheres, ellipsoids, and cylinders); (2) molecular shapes given by the union of overlapping spheres; (3) molecular shapes obtained by exploiting the definition of specific molecular shape functions; (4) isodensity shapes.

In the following we shall limit our analysis to the molecular shapes only, being the latter the most used in the methods we have described in the previous sections. Regarding the procedures corresponding to the item 4 of our classification, namely those giving cavities determined directly from solute isodensity surfaces, we limit to recall that, in their most refined versions they may evaluate the isodensity surface in a self consistent way with the quantum mechanical calculation of the solute wave function [30].

Before going inside the technical details of the single procedures giving cavities of type 2, here represented by the computational technique called GEPOL [31], it is compulsory to make a further classification in terms of the type of molecular surface we have to consider. In general we can define three main kinds of surfaces:

1. the proper van der Waals surface (Sw), which is the external surface resulting from a set of spheres centered on the atoms or group of atoms forming the solute;

2. the surface accessible to the solvent (Sa), defined as the surface generated by the center of the solvent, considered as a rigid sphere, when it rolls around the van der Waals surface;

3. the solvent excluding surface (Se), which can be defined as the contact surface of a probe sphere (with radius equal to the molecular radius of the solvent molecules) rolling on the van der Waals surface.

It is easy to see that Sa can be equivalently defined as a Sw with radii enlarged by a quantity equal to the radius of the solvent, and that the last definition corresponds to the molecular cavity defined by Richards [29].

The choice of the type of surface to be used depends on the specific solute-solvent interactions we want to take into account in the calculation. For example, by exploiting the standard partition of the free energy we can say that steric interactions are better evaluated by employing a Sw cavity, dispersion terms, when derived within the discrete approach, usually require a Sa cavity, while electrostatic terms should exploit a Se surface cavity.

Let us now pass to describe in more details a computational techniques we have chosen among the many developed until now as an example of how a molecular cavity can be defined, and how the tessellation of its surface into tesserae, as needed in all the ASC methods, can be obtained.

GEPOL [31] describes van der Waals (Sa) cavities in terms of spheres centered on solute atoms, and solvent excluding surfaces (Se) by adding some additional spheres. The latter are created with a sequential algorithm: when two spheres of the Sa cavity are close enough to exclude the solvent from the space between them, one or two additional spheres, not centered on atoms, are added, and the procedure is repeated to a prefixed threshold, considering each time all the possible couples of spheres, both of the first and the following generations. The surface of each sphere is then partitioned into triangular tesserae, corresponding to the projection of the faces of a suitable polyhedron inscribed in the sphere. The latter may be defined in different ways, for example a geodesic partition based on pentakisdodecahedron giving origin to 60, 240, or 960 faces can be used (numerical tests have shown that the value of 60 is the best compromise between effectiveness and fastness, and this is the partition most used in the practice). Tesserae with all the vertices inside the cavity are discarded, while those whose surface is partially covered by some other spheres are replaced by the polygonal tesserae with edges on the circles defined by each couple of intersecting spheres. In the most recent version of GEPOL [32] the area a_k of the polygonal tessera k is calculated by applying the Gauss-Bonnet theorem.

6   Derivatives and Molecular Properties

Modern quantum chemistry has been revolutionized by the ability to calculate derivative of the wave function and associated properties with respect to various parameters [33]. Derivatives with respect to variational parameters in the wave function are required in order to find energy stationary points. We also require derivatives of the energy with respect to nuclear positions and externally applied fields to evaluate energy surfaces and molecular response properties, respectively.

The subject of the present section is the direct analytical calculation of (free) energy derivatives from quantum mechanical wave functions taking into account the environment effect, and their application to the evaluation of response properties.

Wave functions contain two kinds of parameters: *explicit* (or *perturbational*) parameters, which have a simple and somewhat arbitrary dependence on the perturbation, such as the basis set origins, and *implicit* (or *variational*) parameters which are the result of a complex computational procedure, such as the already quoted LCAO coefficients. The latter are usually obtained by optimization, either on the energy expression of interest itself or on a simpler energy expression. Let us focus on the explicit parameters, and the way the related analysis is modified by the inclusion of the solvent.

6.1   Free Energy Surfaces

In the already quoted Born-Oppenheimer fixed-nuclei approximation, the total molecular energy is approximated by a sum of the kinetic energy of the nuclei and the molecular energy for fixed (infinitely massive) nuclei. This separation is an excellent approximation for most ground-state systems and also for many excited states. The (fixed nucleus) molecular energy as a function of the nuclear coordinates, i.e. the potential energy hypersurface (PES) $E(\mathbf{R})$ is thus of fundamental importance in understanding and predicting chemical and spectroscopic phenomena.

The characterization of a PES is done in terms of the identification of its stationary points, all defined by a zero gradient value, where here the gradient vector $g(R)$ may be interpreted as the negative of the abstract force acting upon a nuclear configuration represented by R; for a stationary point R_{st}, the gradient vector and hence the forces are zero. The relative location of stationary points along potential energy hypersurfaces has a fundamental role in determining the relative stability of various molecules and the feasibility of interconversion processes among chemical species. Since for all the stationary points the gradient is zero by definition, it is natural to consider the higher derivatives of the potential energy hypersurface for their classification. The second derivatives of $E(R)$ generate the Hessian matrix $H(R)$. The Hessian $H(R_{st})$ of any stationary point is a symmetric real matrix, hence all of its eigenvalues are real; moreover, when the stationary point is a an energy minimum, these eigenvalues are all positive. For stationary points other than minima the Hessian matrix may have one or several negative eigenvalues.

The introduction of solvent effects does not change this kind of analysis; we have only to substitute the potential energy surface with the free energy analog, $G(R)$. From a theoretical point of view, there are no conceptual difficulties to compute the whole portion of free energy hypersurface needed to get all the stationary points taking into account solvent effects. In practice, however, this approach is not viable, due to its large computational cost. In the analogous problem for isolated molecules *in vacuo*, it has been definitely shown that the most effective procedures are those based on analytical expressions of the molecular energy gradient and of the higher-order derivatives.

In the case of molecules in solution, however, the use of analytical derivatives of the free energy with respect to nuclear coordinates is a more complex task. All the continuum methods we have previously considered are characterized by the presence of boundary conditions defining the portion of space where there is no solvent (i.e. the cavity embedding the solute), hence the evaluation of $\partial^n G(R)/(\partial R_x \partial R_y ..)$ must include a reliable calculation of partial derivatives of the boundary conditions. In addition, as we have already shown, the free energy hypersurface is composed of several terms, each depending on solute nuclear coordinates and each giving rise to its own set of partial derivatives with respect to a specific Cartesian coordinate of a solute nucleus.

In the following we shall focus on electrostatic contributions and on the ASC approach only. It can be easily seen that in treating nuclear geometries of molecular solutes immersed in continuous dielectrics, important variations with respect to the isolated system can be expected only in presence of strong solute-solvent interactions, exactly as those due to electrostatic forces in polar or polarizable systems. This does not mean that the other contributions are always negligible, but that in general the electrostatic interactions can be safely assumed as the most effective, and that the analysis in terms of these contributions only is usually sufficient for a quantitative evaluation of the global phenomenon.

Analytical derivatives with respect to nuclear displacements

The standard way [34, 35] followed in quantum chemistry solvation methods to get the free energy derivatives with respect to nuclear displacements consists in differentiating the expression of the free energy which contains all the solvent induced terms as obtained after discretizing the apparent charge into point charges. Numerically, this involves the derivatives of the geometrical parameters defining the tessellation of the cavity (tesserae areas, representative points, normal unit vectors, etc.) with all the problems that this could imply (mostly, a loss of accuracy and stability).

Recently it has been proposed a new procedure which makes the whole problem more tractable [36]. The basic idea is that derivatives with respect to any parameter λ are more conveniently obtained by differentiating the original electrostatic equation

$$-\text{div}(\epsilon(\lambda)\nabla V(\lambda)) = 4\pi\rho(\lambda) \tag{38}$$

before discretizing it and then computing the derivative of the potential as the solution of the derivated equation than to do the job in the other way around, that is to start the derivation from the discretized expressions, as done in the standard scheme. In other words one can always compute the required $\partial V/\partial\lambda$ from

$$-\text{div}\left[\epsilon(\lambda)\nabla\left(\frac{\partial V(\lambda)}{\partial\lambda}\right)\right] = 4\pi\frac{\partial\rho(\lambda)}{\partial\lambda} + \text{div}\left[\frac{\partial\epsilon(\lambda)}{\partial\lambda}\nabla V(\lambda)\right] \tag{39}$$

by solving the equation with the same technique as the original one.

In the quantum mechanical framework used until now, we arrive to the following expression for the first derivative of the electrostatic free energy, G_{el}^{λ}:

$$G_{el}^{\lambda} = \begin{array}{l} tr\mathbf{Ph}^{\lambda} + \frac{1}{2}tr\mathbf{PG}^{\lambda}(\mathbf{P}) - tr\mathbf{S}^{\lambda}\mathbf{PF}'\mathbf{P} + V_{NN}^{\lambda} \\ +D(\partial\rho_M/\partial\lambda,\sigma) + \frac{1}{2}G_{\Sigma}^{\lambda} \end{array} \tag{40}$$

where we have followed the convention which indicates the partial derivatives with upper indices, so that $X^{\lambda} = \partial X/\partial\lambda$. The first four terms are those which compose the energy derivative *in vacuo*, while the Fock matrix in the third term is that given in the presence of the solvent (see 28).

On the other hand, the fifth and sixth terms are new. The former denotes the interaction energy between the variation of the solute charge distribution $\partial\rho_M/\partial\lambda$ and the apparent charge σ:

$$D\left(\frac{\partial\rho_M}{\partial\lambda},\sigma\right) = \int\int_{R^3\times R^3}\left[\frac{\partial\rho_M^N}{\partial\lambda} + \sum\mathbf{P}_{\mu\nu}\frac{\partial\left(\chi_\mu\chi_\nu\right)}{\partial\lambda}\right]\frac{\sigma(y)}{|x-y|}dxdy$$

while the latter represents the only term induced by the changing of the cavity according to motions of the nuclear coordinates; in particular, it can be defined as the derivative of the self-interaction energy of the charge ρ_M, when the cavity is deformed while the charge itself remains fixed.

In practice G_{Σ}^{λ} is computed as a simple integration on all the apparent charges lying on the moving part Σ of the cavity itself, namely:

$$G_{\Sigma}^{\lambda} = \frac{4\pi\epsilon}{\epsilon - 1}\int_{\Sigma}\sigma^2(x)(U_{\Sigma}^{\lambda}(x)\cdot n(x))dx \tag{41}$$

where we denote with U_Σ^j the motion of the interface Σ when λ varies. This quantity assumes a very simple form for standard ASC cavities given as union of spheres, each of them centered on a solute nucleus.

The free energy gradients, thus obtained, can be directly used in standard algorithms for the search of minima in the free energy hypersurface.

It is to be remarked that following the same theoretical scheme it is possible to derive analytical Hessian matrices for solvated systems; here however the technical difficulties are much harder, and for the moment an 'incomplete' scheme, only partially exploiting this simplified procedure, has been computationally implemented [37].

6.2   External Properties

In this section, and more in details in the following subsections, we consider the interactions between molecular systems and external electromagnetic fields. The subject is so large and the related physical phenomena so numerous – suffice it to quote the various spectroscopies – that we shall necessarily limit our exposition to some specific aspects.

In particular, we shall show how to introduce the previous solvation methods into the most used QM formalism exploited to evaluate molecular response properties to electric fields; the parallel treatment of magnetic fields is in fact less interesting from the point of view of this exposition as no specific considerations are required to take into account the solvent effects. In the magnetic case, in fact, the parallelism with the general theory formulated for isolated systems is almost exact, what one needs to do is just to include solvent induced terms into the Hamiltonian of interest. The analysis of external electric fields, on the contrary, requires important modifications when a solvent is also present.

Electric (Hyper)Polarizabilities in solution: an example of dynamic systems

In brief, the global effect of an applied external electrical field on a given molecular system has to be seen in terms of distortions both in the electronic charge distribution (i.e. the field acts as a source of electronic response) and in the nuclear motions which lead to vibrational contributions to (hyper)polarizabilities. However, in the following, we shall limit our analysis to the electronic component only; as concerns the vibrational part, in fact, the solvent does not introduce any specificity and the whole derivation is completely equivalent to that formulated for isolated systems, we only need to include solvent terms in the evaluation of the required quantities.

For the calculation of the electronic (hyper)polarizabilities, there are two conceptually different approaches: the sum-over-states (SOS) and the derivative methods. The SOS method is based on the perturbation expansion of the Stark energy. Different-order Stark energy terms are related to optical nonlinearities on the bases of their order in the field strength. The result is expressions for the polarizability and hyperpolarizabilities as infinite sums over various excited states in which the numerators contain dipole and transition dipole moments between couples of states.

The second method, here indicated as derivative method involves quantum calculations of energy or dipole moment followed by obtaining derivatives either numerically or analytically. The analytical approach for the derivative method involves the coupled Hartree-Fock (CHF) perturbation theory [38].

Following the time-dependent CHF scheme (TD-HF), a term which represents the interaction between the external electric field and the instantaneous molecular dipole is added to the ordinary Hartree-Fock one-electron potential, from (2) it readily follows

$$\hat{H} = \hat{H}^0 + \hat{\rho}_r V_r^R + \hat{\rho}_r \mathbf{V}_{rr'}^R \langle \Psi | \hat{\rho}_{r'} | \Psi \rangle + \sum_k^{Ne} \hat{\mu}(k) \cdot E_{ex} \tag{42}$$

where $\hat{\mu}(k)$ the dipole moment operator associated to the kth electron of the solute. In the most general case of both static and oscillating monochromatic fields the expression to be used is $E_{ex} = E(e^{+i\omega t} + e^{-i\omega t} + 1)$.

Approximate solutions of the time-dependent Schrödinger equation associated to the effective Hamiltonian (42) can be obtained by using Frenkel variational principle [39], which in the solvation method can be cast in the form [40]

$$\delta \left\langle \Psi \left| \hat{\mathcal{G}} - i\frac{\partial}{\partial t} \right| \Psi \right\rangle = 0 \quad \text{with } \frac{\partial}{\partial t} \langle \delta\Psi | \Psi \rangle = 0 \tag{43}$$

In (43) $\hat{\mathcal{G}}$ represents the free energy operator corresponding to the functional defined in (1). In the static limit the problem is reduced to a time-independent Schrödinger equation whose variational solution coincides with the stationary condition on $\mathcal{G}$.

The restriction to a one-determinant wavefunction (N electrons, closed shell case) with orbital expansion over a finite basis set leads to the following time-dependent Hartree-Fock equation for solvated systems [41]:

$$\left(\mathbf{F}' - i\frac{\partial}{\partial t} \right) \mathbf{T} = \mathbf{T}\epsilon \tag{44}$$

$$\mathbf{F}' = \mathbf{h} + \mathbf{h}^R + \sum_a \mathbf{m}_a E_{ex}^a + \mathbf{G}(\mathbf{P}) + \mathbf{X}^R(\mathbf{P}) \tag{45}$$

Here $\mathbf{m}^a$ collects the integrals of the a-th Cartesian component of the dipole moment operator.

In the CPHF approach the matrices of (44) are expanded in powers of the external field components and the resulting equation is partitioned following the orders of differentiation into a set of Hartree-Fock equations. This scheme allows to obtain the various electric properties (dipole, polarizability and hyperpolarizabilities) in terms of the density matrix for the unperturbed system ($\mathbf{P}_0$) and its variations ($\mathbf{P}^{abc..}$, at the various orders of differentiation) due to the presence of the field [42]:

$$\begin{aligned}
\mu_a^0 &= -tr[\mathbf{m}_a \mathbf{P}_0] \\
\alpha_{ab} &= -tr[\mathbf{m}_a \mathbf{P}^b] \\
\beta_{abc} &= -tr[\mathbf{m}_a \mathbf{P}^{bc}] \\
\gamma_{abcd} &= -tr[\mathbf{m}_a \mathbf{P}^{bcd}]
\end{aligned} \tag{46}$$

Until now an exact parallelism between the present solvation model and the gas-phase theory is obtained as the definition of the solvent operators $\mathbf{h}^R$ and $\mathbf{X}^R$ do not introduce any further difficulty in the theory originally formulated for isolated systems.

However, if we go more deeply inside the problem, new solvent-specific considerations have to be taken into account. When oscillating fields are present, in fact, a new dynamic description of the solvent effects has to be introduced. Here, the analysis will necessarily be restricted to the electrostatic interaction only as the inclusion of the other two contributions, repulsion and dispersion, would imply deep reformulations of their theoretical bases which have not been defined yet.

In brief, the study of the response of a molecular solute to a time dependent perturbation requires a non-equilibrium formulation of the solvation method. The introduction of the dynamical aspect into methods accounting for solvation effects has a long history [43]; one of the most successful theory identifies the solvent response function with a polarization vector depending on time, $P(t)$, which varies according to the variation of solute charge distribution. This vector accounts for many phenomena related to different physical processes taking place inside and among the solvent molecules. In practice strong approximations are necessary in order to formulate a feasible model to deal with such complex quantity. A very simple but also effective way of seeing this problem is to partition the polarization vector in terms accounting for the main degrees of freedom of the solvent molecules, each one determined by a proper relaxation time (or by the related frequency).

In most applications, this spectral decomposition is limited to two terms only representing "fast" and "slow" phenomena, respectively

$$P(t) = P_{fast} + P_{slow} \tag{47}$$

In particular, the fast term is easily associated with the polarization due to the bound electrons of the solvent molecules which can instantaneously adjust themselves to any change of the solute charge distribution. The slow term on the contrary, even if it is often referred to as a general orientational polarization, has a less definite nature. Roughly speaking, it collects many different nuclear and molecular motions (vibrational relaxations, rotational and translational diffusion, etc.) related to generally much longer times with respect to those involved in the changes of the solute description and/or of the electronic polarization. Only when both the fast and the slow terms are adjusted to the actual solute description we have solvation states corresponding to full equilibrium. The more general nonequilibrium solvation states are, on the contrary, characterized by the quantum electronic state of the solute and the solvent electronic polarization appropriate to an arbitrary nonequilibrium solvent slow polarization.

This kind of analysis is easily shifted to ASC models in which the polarization vector is substituted by the apparent surface charge in the representation of the solvent reaction field. In this framework, the previous partition of the polarization vector into *fast* and *slow* components leads to two corresponding surface apparent charges, σ_f and σ_s, the sum of which gives the total apparent charge σ. The definition of these charges is in turn related to the static dielectric constant $\epsilon(0)$ (for the full equilibrium total charge σ), to the frequency dependent dielectric constant $\epsilon(\omega)$ (for the fast component σ_f), and to a combination of them (for the remaining slow component σ_s).

In the TD-SCF scheme above this is reflected in the fact that the effective solvent contribution determined by the interaction matrix $\mathbf{X}(\mathbf{P})$ (see 45) in the presence of an oscillating perturbation has to be limited to the fast charge only. In this way it is thus possible to take into account all the needed details of the distribution of $\varepsilon = \varepsilon(\omega_\sigma)$ within the range of frequencies associated to the resulting external field (if only a static field is present ε reduces to the static dielectric constant). If, for example, we assume that the solvent is polar and follows a Debye-like behavior, we have the general relation:

$$\epsilon(\omega) = \epsilon(\infty) + \frac{\epsilon(0) - \epsilon(\infty)}{1 - i\omega\tau_D}$$

with $\epsilon(0) = 78.5$, $\epsilon(\infty) = 1,7756$, and $\tau_D = 0.85 \cdot 10^{-11} s$ in the specific case of water.

It is clear that at the optical frequencies, the value of $\epsilon(\omega)$ is practically equal to $\epsilon(\infty)$; variations with respect to this, in fact, become important only at $\omega \ll 1/\tau_D$ where $\epsilon(\omega)$ is close to the static value. In practice, this means that for polar solvent like water, where $\epsilon(\infty)$ is much smaller than the static value, the solvent effects are much more effective when all the solvent degrees of freedom are equilibrated to the solute than in the case of only a partial equilibrium of its fastest components. Numerically this means that response properties computed in the full solute-solvent equilibrium or, instead in a condition of nonequilibrium will present quite different values.

This important modification of standard solvation models so that to take into account dynamical effects, can be extended to many different processes, here we limit ourselves to recall vertical electronic transitions. According to the Franck-Condon principle, the spectrum (either of absorption and of fluorescence) should be calculated without allowing nuclear motion, i.e. in our scheme, using a solvent reaction field in which the slow part is equilibrated to the ground state (or the excited state in case of fluorescence), but in which the electronic structure of the solute and solvent are optimized for the excited state (or viceversa the ground state) [44].

7   An example of applications

In principle, solvation methods based on effective Hamiltonian approaches allow all of the analyses and the manipulations of the solute wave functions that are possible in gas phase. Presently, the most advanced computational procedures, based on the methods illustrated in the preceding sections, can provide:

- the solute molecular free energy, partitioned into electrostatic and non electrostatic terms: the *ab initio* electrostatic contributions can be calculated for closed- and open-shell systems at Hartree-Fock (HF) and Density Functional

Theory (DFT) levels, and at some post-HF levels (i.e. MP2, MP4, Coupled Cluster, CI, CAS-SCF);

- the energy gradients with respect to nuclear coordinates, to be used for geometry optimizations in condensed phase;

- the evaluation of harmonic force constants;

- the calculation of first order electronic properties (dipole moments, Mulliken or Bader atomic charges, spin densities and so on);

- the calculation of molecular response properties (i.e. the response of the molecular solute to external electromagnetic fields) and the evaluation of electronic and vibrational spectra.

The above quantities can be used to study the structure, the properties and the reactivity of chemical systems in solution, following the same procedures adopted for isolated molecules.

In the following we report an example describing two specific applications extracted from the list above. The goal here, in fact, is not to give numerical proofs of all the aspects that can be computationally treated but to provide some examples of how theoretical models are usually applied to study real chemical systems. We refer the interested reader to the reviews reported in ref. [1] for an extended view on applications of theoretical solvation models and on the corresponding large literature.

The results we summarize below have been extracted from a study on the electric response properties of molecular systems in solution published on the Journal of the American Society [47]. This study has been performed using one specific solvation method we have described in the previous section, namely the ASC method called IEF. Within this framework, the molecular solute is embedded in a cavity in the dielectric medium defined in terms of interlocking spheres centered on the solute nuclei, with radii R_k equal to 1.2 times the corresponding van der Waals values R_k^{vdw}. The calculations have been performed exploiting development versions of two largely diffused computational codes (Gaussian [45] and GAMESS [46]) in which the IEF method has been implemented.

7.1 Structures and (hyper)polarizabilities of push-pull systems in solution

In this subsection we summarize some numerical results on geometries and static electronic (hyper)polarizabilities of two series of noncentrosymmetric polyenes, $NH_2(CH=CH)_N R$ (N=1,2), with R=CHO (series I) and with R=NO_2 (series II), in vacuo and in aqueous solution.

These systems belong to the category of push-pull π-conjugated molecules, i.e. systems where a conjugated linker segment is capped by a donor group on one end and an acceptor on the other. Many donors and acceptors, as well as conjugated linkers of various nature and length have been investigated both theoretically and experimentally [48]. The specificity of this kind of molecules, in particularly we shall focus on substituted polyenes, is that their ground-state (GS) structure can be represented as a combination of different resonance forms, differing in the extent of charge separation.

(Polyene-like)   (Cyanine-like)   (Polymethine-like)

When the donor and acceptor groups are weak, the neutral polyene-like resonance form dominates the ground state and the molecule has a structure with a distinct alternation in the bond length between neighboring carbon atoms, i.e. a high degree of bond length alternation (BLA), defined as the difference between average single an double bond distances in the conjugated pathway. When donor and acceptor substituents become stronger, the contribution of ionic resonance forms to the ground state increases, and BLA first decreases until almost zero values (in the intermediate cyanine-like structure with partial charge separation) and then increases again towards the high negative values proper of the fully charge-separated polymethine-like (or zwitterionic) form. The relative contribution of these three resonance forms to the GS is also controlled by the polarity of the solvent in which the chromophore is dissolved; a more polar solvent increases the GS state polarization, which makes the partial and/or complete ionic forms more important.

Experiments have clearly demonstrated that the medium influences both the molecular geometry and the electronic charge distribution leading thus to large effects on electronic linear and nonlinear optical properties [48].

All the calculations have been performed at SCF level with a basis set equal to the Dunning/Huzinaga valence double-zeta [49]. The solvation calculations are performed for a medium having dielectric constant $\epsilon = 78.5$ corresponding to the static dielectric constants of liquid water at 298 K.

In Table 1 we report the results obtained for the static electronic polarizability and hyperpolarizability for the two series of molecules both in vacuo and in solution. As the molecules are taken to lie in the xy-plane (actually the exact planarity is lost in solution) with the long axis oriented along the x-axis, the dominant components are α_{xx} for the polarizability, and β_{xxx} for the first hyperpolarizability, in the following our analysis will be limited to these diagonal components only.

Table 1. Static electronic (hyper)polarizabilities (a.u.) of $H_2N(CH=CH)_N R$ (with R=CHO,NO$_2$ and N=1,2) in vacuo and in water.

N	CHO		NO$_2$	
	1	2	1	2
vacuum				
α_{xx}^e	70.6	140.5	71.3	146.7
β_{xxx}^e	-398.7	-1437.6	-490.4	-1893.7
water				
α_{xx}^e	103.0	231.2	122.0	319.4
β_{xxx}^e	-899.9	-4701.3	-1227.2	-1928.4

To make our analysis clearer, the discussion on the results will be divided in two parts, focussed on geometry and electronic effects, respectively.

Geometry Effects

As reported above, the extent of geometric distortions induced by solvent or an external electric field can be characterized by bond-length alternation (BLA), defined as the average difference in length between single and double bonds in the conjugated pathway (it is by convention taken as positive in the neutral polyene-like form).

In Tables 2 and 3 we report the bond lengths of the first element of both series at the geometries optimized in vacuo and in solution, while in Table 4 the BLA values for the second terms of each series of molecules (for the first terms it is not possible to define a BLA as the conjugated path is limited).

Table 2. Bond lengths (Å) of H$_2$N(CH=CH)R in vacuo and in water; if R=CHO X=C and Y=O, otherwise, when R=NO$_2$, X=N and Y=O.

	CHO		NO$_2$	
	vacuum	water	vacuum	water
R_{XY}	1.2284	1.2506	1.2478	1.2696
R_{CX}	1.4543	1.4304	1.4216	1.3721
R_{CC}	1.3511	1.3696	1.3494	1.3809
R_{CN}	1.3664	1.3448	1.3552	1.3236

Table 3. Bond lengths (Å) of H$_2$N(C$_d$H=C$_c$H-C$_b$H=C$_a$H)R in vacuo and in water; if R=CHO X=C and Y=O, otherwise, when R=NO$_2$, X=N and Y=O.

	CHO		NO$_2$	
	vacuum	water	vacuum	water
R_{XY}	1.2274	1.2477	1.2464	1.2830
R_{C_aX}	1.4608	1.4385	1.4311	1.3540
$R_{C_aC_b}$	1.3483	1.3645	1.3440	1.3949
$R_{C_bC_c}$	1.4507	1.4329	1.4413	1.3896
$R_{C_cC_d}$	1.3499	1.3661	1.3550	1.3966
R_{CN}	1.3728	1.3526	1.3644	1.3229

Table 4. Bond length alternation (BLA) (Å) of $H_2N(CH=CH)_2R$ in vacuo and in water.

	vacuum	water
CHO	0.1067	0.0704
NO_2	0.0917	-0.00614

The two main aspects to be pointed out are the effects due to the nature of substituents and, more important, to the presence of the solvent.

Both Tables 2 and 3 show that more polar D-A substituents (i.e. molecules of series II) act to reduce the length of the single bonds and to increase the length of the double bonds. The same effect, largely amplified, is given by the presence of the solvent; in the latter case the BLA decreases from 0.1067 to 0.0704 for $NH_2(CH=CH)_2CHO$, and from 0.0917 to -0.0061 for $NH_2(CH=CH)_2NO_2$, passing from gas phase to aqueous solution.

In addition, data reported in Table 4 show that, as one might expect, the presence of a polar solvent like water induces a migration of electronic charge distribution along the molecular axis from D to A group; in this way the electronic structure of the molecular system, which in the gas phase is well characterized by a polyene-like structure only, in solution contains a large contribution also from the partially charge-separated cyanine-like form: in $NH_2(CH=CH)_2NO_2$ in water the BLA becomes almost zero, exactly as in an ideal cyanine structure.

The same conclusions can be reached in a different way, by taking into account the changes on the net charges of the various chemical groups present in the molecule, when passing from gas phase to solution. In Table 5 we report the total dipole moment and the Mulliken net charges on the D-A groups as well as on the intermediate CH groups both in vacuo and in solution.

Table 5. Ground state dipole moment (debye) and Mulliken net charges (a.u.) of $R-(X1=X2-X3=X4)-NH_2$ (with R=CHO,NO_2) both in gas phase and in aqueous solution.

	μ_x	group charges (a.u.)					
		R	X1	X2	X3	X4	NH_2
$CHO_{(g)}$	-6.3179	-0.1555	-0.0531	0.2699			-0.0612
$CHO_{(aq)}$	-9.2531	-0.2602	-0.1131	0.3475			0.0258
$CHO_{(g)}$	-7.7909	-0.1605	-0.0571	0.1037	-0.0381	0.2339	-0.0820
$CHO_{(aq)}$	-12.0932	-0.2599	-0.1253	0.1769	-0.0935	0.3047	-0.0029
$NO_{2(g)}$	-8.1474	-0.4420	0.0705	0.3925			-0.0211
$NO_{2(aq)}$	-12.3971	-0.6449	0.0684	0.4582			0.1183
$NO_{2(g)}$	-10.1829	-0.4492	0.0514	0.2323	-0.0569	0.2815	-0.0590
$NO_{2(aq)}$	-19.6363	-0.7548	0.0296	0.2807	-0.0656	0.4059	0.1041

The most evident result of data reported in Table 5 is that the A group which in vacuo shows significative partial negative charges in both series, in solution present a positive charge, with the only exception of the second term of series I where the

A net charge is almost zero. This behavior once again shows that in solution the cyanine-like structure assumes a much more important role and becomes the main one for the series II. The electron-migration towards the A group is reflected also in the net charges of the intermediate CH groups which all increase in magnitude passing from vacuum to aqueous solution.

Electronic effects

The larger mobility of the electronic charge in the presence of the solvent we have derived from the analysis on geometrical changes of previous subsection is also made evident by the significative increase in the electronic polarizability passing from gas phase to solution. This is shown in Table 6 in which we report the percent variation of electronic polarizability and first hyperpolarizability of both series of molecules.

Table 6. Percent variations of static (hyper)polarizabilities of $H_2N(CH=CH)_NR$ passing from the gas phase to the aqueous solution: $\Delta M = 100*[M(aq)-M(gas)]/M(gas)$

	CHO		NO$_2$	
N	1	2	1	2
$\delta\alpha^e_{xx}$	+46	+64	+71	+118
$\delta\beta^e_{xxx}$	+126	+227	+150	+2

The first hyperpolarizability β^e which, in both terms of the series I, presents significative increments going from gas phase to solution (from 126% to 227%), in the series II shows an unexpected behavior with a very small increase in the solvated $NH_2(CH=CH)_2NO_2$ molecule; here β^e is only 2% larger than the vacuum result. This result, which is completely different from those of the other computed molecules and also from the usual values reported in the literature for other push-pull systems, needs a more detailed analysis. In Table 7 we report the Random Phase Approximation (RPA) results of the lowest allowed single excitation from the GS for all the molecules both in vacuo and in solution.

Table 7. Dipole variations (debye) with respect to the GS value, $\Delta\mu^{gn}_x$, transition energies (eV), ΔE^{gn}, and oscillatory strength (a.u.), f, for the lowest allowed single excitation of $H_2N(CH=CH)_NR$ molecules both in gas phase and in aqueous solution.

	f	$\Delta\mu^{gn}_r$	ΔE^{gn}
$CHO(1)_{(g)}$	0.784	-2.017	6.280
$CHO(1)_{(aq)}$	0.989	-1.366	5.369
$CHO(2)_{(g)}$	1.283	-3.446	5.257
$CHO(2)_{(aq)}$	1.523	-2.522	4.154
$NO_2(1)_{(g)}$	0.445	-4.040	5.593
$NO_2(1)_{(aq)}$	0.725	-2.331	4.364
$NO_2(2)_{(g)}$	0.967	-5.448	4.834
$NO_2(2)_{(aq)}$	1.284	-0.859	3.287

The RPA procedure, even if formally equivalent to CPHF, allows one to perform an analysis of the electronic (hyper)polarizabilities in terms of sums of contributions from different electronic excited states (SOS). A very simple scheme which is very useful for interpretative purposes, even if not reliable from a quantitative point of view, limits the SOS expression to a unique excited state. In the resulting two-state approximation (TSA), the static diagonal electronic polarizability and first hyperpolarizability are given by:

$$\alpha_{rr}^e = 2\frac{(\mu_r^{gn})^2}{\Delta E^{gn}} = 3\frac{f^{gn}}{(\Delta E^{gn})^2} \tag{48}$$

$$\beta_{rrr}^e = 4\frac{(\mu_r^{gn})^2 \Delta\mu_r^{gn}}{(\Delta E^{gn})^2} = 6\frac{f^{gn}\Delta\mu_r^{gn}}{(\Delta E^{gn})^3} = 2\frac{\alpha_{rr}^e \Delta\mu_r^{gn}}{\Delta E^{gn}} \tag{49}$$

where μ_r^{gn} is the r component of dipole transition moment between the ground and excited state, ΔE^{gn} is the corresponding excitation energy, $\Delta\mu_r^{gn}$ is the change in the r component of the dipole moment between the ground and the excited state, and $f^{gn} = 2\mu_{gn}^2 \Delta E^{gn}/3$ is the oscillatory strength (in the equations above we have assumed that the transition dipole moment is completely described by its r component, i.e. $\mu_{gn}^2 \simeq (\mu_r^{gn})^2$.

In Table 8 we report the static (hyper)polarizability values obtained both in vacuo and in solution by applying (48-49).

Table 8. Static electronic polarizability and first polarizability (a.u.) in the Two-level model for $H_2N(CH=CH)_N R$ both in the gas phase and in aqueous solution. Values in parentheses refer to the percentage of the property with respect to its CPHF total value.

	α_{xx}^e	β_{xxx}^e
$CHO(1)_{(g)}$	39.22 (55.6)	-269.7 (67.6)
$CHO(1)_{(aq)}$	69.98 (67.9)	-381.2 (42.4)
$CHO(2)_{(g)}$	94.03 (66.9)	-1319.8 (91.8)
$CHO(2)_{(aq)}$	182.2 (78.8)	-2368.6 (50.4)
$NO_2(1)_{(g)}$	31.47 (44.1)	-487.0 (99.0)
$NO_2(1)_{(aq)}$	84.26 (69.0)	-964.3 (78.6)
$NO_2(2)_{(g)}$	91.38 (62.3)	-2206.5 (116.5)
$NO_2(2)_{(aq)}$	263.48 (82.5)	-1475.2 (76.5)

The main point to be stressed is that the TSA only gives qualitative results; the very small discrepancies between TSA and analytical α^e and β^e values found for some molecules are almost fortuitous; in one case TSA even overestimates the analytical β^e. Anyway, TSA establishes an important link between electronic properties and spectroscopic quantities.

¿From data of Table 7, one sees that the significant increase of the polarizability in solution is largely due to the red-shift of the transition energy; in both the series the $\delta\Delta E^{gn}$ is around 1 eV (~ 8000 cm^{-1}). This bathochromic shift, which has been experimentally observed in similar push-pull systems [50], corresponds to a situation where, through the interaction with the solvent reaction field, the first excited state is preferentially stabilized with respect to the GS being its dipole moment larger than that in the GS. Once again, due to the strong link present in these π-conjugated systems between geometry and electronic structure, the same conclusions can be reached from considerations on BLA changes: the red-shift in the transition can take place as the GS in solution becomes described by equal contributions from the polyene form and the partially charge-separated form, i.e. the cyanine limit (and in fact the largest shift, $\delta\Delta E^{gn} = 1.54$ eV, is given by $NH_2(CH=CH)_2NO_2$ for which a very small BLA is also found).

In the RPA-SOS framework also the unusual behavior found for the β^e of $NH_2(CH=CH)_2NO_2$ in solution assumes a clearer meaning. From what reported in Table 7, it comes out that the most irregular datum regarding this molecule is the very small change in the transition dipole moment from the GS to the first allowed excited state in solution with respect to the value in vacuo (-0.859 debye vs -5.448 debye); due to the proportionality of β^e and $\Delta\mu_x^{gn}$, the final result is the observed small increase of the first hyperpolarizability of the solvated system with respect to the value computed in vacuo. On the contrary, this kind of behavior is not shown by the polarizability which is not related to $\Delta\mu_x^{gn}$; as a matter of fact, α^e presents a standard large increase also for $NH_2(CH=CH)_2NO_2$.

To better understand this phenomenon we recall that the excited state can be almost exclusively described by one basic excitation (typically highest occupied molecular orbital, HOMO $\rightarrow$ lowest unoccupied molecular orbital, LUMO). For all these conjugated molecules the HOMO-LUMO transition is a $\pi \rightarrow \pi^*$ transition. The general form of the HOMO in $NH_2(CH=CH)_2NO_2$ does not change too much from gas phase to solution, and in fact the coefficients, $\{c_i\}$ of the two MOs relatively to the atomic functions centered on the two N atoms of the D-A groups are very similar ($q_{N_D} = \sum_i c_i^2(N_D)$ is around 0.1 and q_{N_A} is around 0.0 in both cases). Things are different for the LUMO, here the effective charge q_{N_D} on the N of the amine changes from 0.070 to 0.116 passing from vacuo to solution. This means that in the charge-transfer excitation HOMO-LUMO, the variation of the q_{N_D} is -0.108 in vacuo and almost zero in water (the parallel value of δq_{N_A} is 0.253 in vacuo and 0.269 in solution); consequently, the dipole moment change passing from the GS to the excited state, which in vacuo presents a large value, is very small in solution, as small is the increase of β^e.

8   Conclusions

As said in the Introduction, the theoretical representation of condensed phases can now be achieved by a large variety of techniques; the review given in this work necessarily represents a limited portion of the real 'panorama' but it still should manage to give the 'flavor' of this research field. Numerous reviews and original papers on the argument have been quoted but many other could be found; here, however, the main scope was not to give a complete and well documented summary of the past and the present of the quantum mechanical approaches to the problem of condensed phases.

The initial idea was to describe the main theoretical aspects of a specific solvation approach which, we think, presents many features of some interest also for scientists not strictly related to the world of the chemical research. The large freedom in the definition of the formal strategy to follow and the many numerical manipulations involved in its computational implementation should in fact represent attractive features for different types of theoreticians, and in particular for mathematicians.

We are sure that, due to the relative youth of this field, important new features, in particular as concerns the development of the formal aspects of the problem, can be still introduced, and that completely new ideas, eventually revolutionizing the present trend of the research, can be suggested without the risk of being rejected as it often happens in other more stated areas of the scientific research.

The few past examples of joint projects among groups belonging to different fields show that the contribution from other scientific disciplines is of great help for the progress of the theoretical chemistry in general, and, more in particular, of the theoretical modellization of condensed phases. We hope that the present paper will contribute to increase this still too small number of synergic experiences by suggesting further points of common interest or at least by switching on the curiosity toward the complex field of the interactions between a molecular system and its surrounding.

Bibliography

[1] (a) J. Tomasi and M. Persico, *Chem. Rev.* **94**, 2027 (1997); (b) J.-L. Rivail and D. Rinaldi, in *Computational Chemistry Review of Current Trends*, J. Leszczynski (ed.) (World Scientific, New York, 1995); (c) M. Orozco, C. Alhambra, X. Barril, J.M. Lopez, M.A. Busquest, and F.J. Luque, *J. Mol. Model.* **2**, 1 (1996); C.J. Cramer and D.G. Truhlar, *Chem. Rev.* **99**, 2161 (1999).

[2] S. Yomosa, *J. Phys. Soc. Jap.* **35**, 1738 (1973).

[3] D. Rinaldi and J.-L. Rivail, *Theor. Chim. Acta* **32**, 57 (1973).

[4] J.E. Sanhueza, O. Tapia, W.G. Laidlaw, and M. Trsic, *J. Chem. Phys.* **70**, 3096 (1979).

[5] H. Hoshi, M. Sakurai, Y. Inoue, and M. Chujo, *J. Chem. Phys.* **87**, 1107 (1987).

[6] A. Ben-Naim, (a) *Water and Aqueous Solutions*, Plenum Press. New York 1974; (b) *Solvation Thermodinamics*, Plenum Press, New York, 1987.

[7] H.H. Uhlig, *J. Phys. Chem.* **41**, 1215 (1937).

[8] O. Sinanoglu, *J. Chem. Phys.* **75**, 463 (1981).

[9] V. Gogonea and E. Osawa, *J. Mol. Struct. (Theochem)* **311**, 305 (1994).

[10] (a) H. Reiss, H.L. Frisch and J.L. Lebowitz, *J. Chem. Phys.* **31**, 369 (1959); (b) R. A. Pierotti, *Chem. Rev.* **76**, 717 (1976); (c) J. Langlet, P. Claverie, J. Caillet and A. Pullman, *J. Phys. Chem.* **92**, 1617 (1988).

[11] (a) F.M. Floris and J. Tomasi, *J. Comput. Chem.* **10**, 616 (1989); (b) F.M. Floris, J. Tomasi and J.L. Pascual-Ahuir, *J. Comput. Chem.* **12**, 784 (1991).

[12] C. Amovilli and B. Mennucci, *J. Phys. Chem. B* **101**, 1051 (1997).

[13] E. Dzyaloshinsky, E.M. Lifchitz and L.P. Pitaevskii, *Adv. Phys.* **10**, 165 (1961).

[14] B. Linder, *Adv. Chem. Phys.* **12**, 225 (1967).

[15] D. Rinaldi, B.J. Costa-Cabral and J.L. Rivail, *Chem. Phys. Lett.* **125**, 495 (1986).

[16] M. A. Aguilar and F.J. Olivares del Valle, *Chem. Phys.* **138**, 327 (1989).

[17] S. Miertuš, E. Scrocco and J. Tomasi, *Chem. Phys.* **55**, 117 (1981).

[18] R. Cammi and J. Tomasi, *J. Comp. Chem.* **16**, 1449 (1995).

[19] A. Klamt and G. Schüürmann, J. *Chem. Soc. Perkin Trans.* **2**, 799 (1993).

[20] T. N. Truong and E.V. Stefanovich, *Chem. Phys. Lett.* **240**, 253 (1995).

[21] (a) E. Cancès and B. Mennucci, *J. Math. Chem.* **23**, 309 (1998); (b) E. Cancès, B. Mennucci and J. Tomasi, *J. Chem. Phys.* **107**, 3032 (1997); (c) B. Mennucci, E Cancès and J. Tomasi, *J. Phys. Chem. B* **101**, 10506 (1997).

[22] (a) D. Rinaldi, M.F. Ruiz-López and J.L. Rivail, J. Chem. Phys. **78**, 834 (1983); (b) V. Dillet, D. Rinaldi and J.L. Rivail, J. Phys. Chem. **98**, 5034 (1994).

[23] C.J. Cramer, G.C. Lynch, G.D. Hawking, D.G. Truhlar, QCPE Bull: AMSOL version 4.0 13, 78 (1993).

[24] C.J. Cramer and D.G. Truhlar, Reviews on Computational Chemistry, D.B. Boyd and K.B. Lipkowitz (eds.), VCH, New York (1995); Vol. 6.

[25] V. Dillet, D. Rinaldi, J. Bertrán and J.L. Rivail, J. Chem. Phys. **104**, 9437 (1996).

[26] E. Cancès, C. Le Bris, B. Mennucci and J. Tomasi, Mathematical Models and Methods in Applied Sciences, **9**, 35 (1999).

[27] D. E. Beskos, *Boundary Element Methods in Mechanics* (North Holland: Amsterdam, 1997).

[28] W. Hackbusch, *Integral Equations–Theory and Numerical Treatment* (Birkhäuser Verlag: Basel, Switzerland, 1995).

[29] B. Lee and F.M. Richards, J. Mol. Biol. **55**, 379 (1971).

[30] J.B. Foresman, T.A. Keith, K.B. Wiberg, J. Snoonian, M.J. Frisch, J. Phys. Chem. **100**, 16098 (1996).

[31] (a) J.L. Pascual-Ahuir, E. Silla, J. Tomasi and R. Bonaccorsi, *J. Comp. Chem.* **8**, 778 (1987); (b) J.L. Pascual-Ahuir, E. Silla, *J. Comp. Chem.* **11**, 1047 (1990); (c) J.L. Pascual-Ahuir, E. Silla, I. Tuñón, *J. Comp. Chem.* **15**, 1127 (1994).

[32] M. Cossi, M. Mennucci and R. Cammi, J. Comp. Chem. **17**, 57 (1996).

[33] (a) P. Jorgensen and J. Simons, *Geometrical Derivatives of Energy Surfaces and Molecular Properties* (Reidel, Dordrecht, 1986); (b) P. Pulay, Adv. Chem. Phys. **69**, 241 (1987); (c) Y. Yamaguchi, Y. Osamura, J.D. Goddard and H.F. Schaefer III, *A New Dimension to Quantum Chemistry: Analytic Derivative Methods in Ab Initio Molecular Electronic Structure Theory* (Oxford University Press, Oxford, 1994); (d) P. Pulay, *Analytical Derivative Techniques and the Calculation of Vibrational Spectra*, in *Modern Electronic Structure Theory, Part II*, D.R. Yarkony (ed.), (World Scientific, Singapore 1995).

[34] (a) R. Cammi and J. Tomasi, J. Chem. Phys. **100**, 7495 (1994); (b) R. Cammi and J. Tomasi, J. Chem. Phys. **101**, 3888 (1994).

[35] (a) T.N. Truong and E.V. Stefanovich, *J. Chem. Phys.* **103**, 3709 (1995); (b) E.V. Stefanovich and T.N. Truong, *J. Chem. Phys.* **105**, 2961 (1996); (c) J. Andzelm, C. Kolmel and A. Klamt, *J. Chem. Phys.* **103**, 9312 (1995).

[36] (a) E. Cancès and B. Mennucci, *J. Chem. Phys.* **109**, 249 (1998); (b) E. Cancès, B. Mennucci and J. Tomasi, *J. Chem. Phys.* **109**, 260 (1998).

[37] B. Mennucci, R. Cammi, J. Tomasi, *J. Chem. Phys.*, **110**, 6858 (1999).

[38] R. McWeeny, *Methods of Molecular Quantum Mechanics* (Academic Press, London, 1992); second edition.

[39] J. Frenkel, *Wave Mechanics-Advanced General Theory* (Oxford Univ. Press, London 1989).

[40] R. Cammi and J. Tomasi, Int. *J. Quantum Chem.* **60**, 1165 (1996).

[41] R. Cammi, M. Cossi, B. Mennucci and J. Tomasi, *J. Chem. Phys.* **105**, 10556 (1996).

[42] (a) H. Sekino and R.J. Bartlett, *J. Chem. Phys.* **85**, 976 (1986); (b) S.P. Karna and M. Dupuis, *J. Comp. Chem.* **12**, 487 (1991).

[43] (a) Y. Ooshika, *J. Phys. Soc. Jpn.* **9**, 594 (1954); (b) R. Marcus, *J. Chem. Phys.* **24**, 966 (1956); (c) E.G. McRae, *J. Phys. Chem.* **61**, 562 (1957); (d) S. Basu, *Adv. Quantum. Chem.* **1**, 145 (1964); (e) V.G. Levich, *Adv. Electrochem. Eng.* **4**, 249 (1966) for older studies and (f) H.J. Kim, and J.T. Hynes, *J. Chem. Phys.* **93**, 5194 and 5211 (1990); (g) A.M. Berezhkosvskii, *Chem. Phys.* **164**, 331 (1992); (h) M.V. Basilevski, and G.E. Chudinov, *Chem. Phys.* **144**, 155 (1990); (i) M.A. Aguilar, F.J. Olivares del Valle, and J. Tomasi, *J. Chem. Phys.* **98**, 7375 (1993); (l) R. Cammi, J. Tomasi, *Int. J. Quantum Chem: Quantum Chem. Symp.* **29**, 465 (1995); (m) K.V. Mikkelsen, A. Cesar, H. Ågren, and H. J. Aa. Jensen, *J. Chem. Phys.* **103**, 9010 (1995) for more recent ones.

[44] (a) M.L. Sánchez, M.A. Aguilar and F.J. Olivares del Valle, *J. Phys. Chem.* **99**, 15758 (1995); (b) B. Mennucci, R.Cammi and J. Tomasi, *J. Chem. Phys.* **109**, 2798 (1998); (c) B. Mennucci, A. Toniolo and C. Cappelli, *J. Chem. Phys.* **110**, 6858 (1999).

[45] M. J. Frisch, G. W. Trucks, H. B. Schlegel, G. E. Scuseria, M. A. Robb, J. R. Cheeseman, V. G. Zakrzewski, J. A. Montgomery, R. E. Stratmann, J. C. Burant, S. Dapprich, J. M. Millam, A. D. Daniels, K. N. Kudin, M. C. Strain, O. Farkas, J. Tomasi, V. Barone, M. Cossi, R. Cammi, B. Mennucci, C. Pomelli, C. Adamo, C S. Clifford, J. Ochterski, G. A. Petersson, P. Y. Ayala, Q. Cui, K. Morokuma, D. K. Malick, A. D. Rabuck, K. Raghavachari, J. B. Foresman, J. V. Ortiz, A. G. Baboul, J. Cioslowski, B. B. Stefanov, G. Liu, C A. Liashenko,

P. Piskorz, I. Komaromi, R. Gomperts, R. L. Martin, D. J. Fox, T. Keith, M. A. Al-Laham, C. Y. Peng, A. Nanayakkara, C. Gonzalez, M. Challacombe, P.M.W. Gill, B. Johnson, W. Chen, C M. W. Wong, J. L. Andres, M. Head-Gordon, E. S. Replogle and J. A. Pople (Gaussian, Inc., Pittsburgh, PA, 1998); Gaussian 99, Development Version.

[46] M.W. Schmidt, K.K. Baldridge, J.A. Boatz, S.T. Elbert, M.S. Gordon, J.H. Jensen, S. Koseki, N. Matsunaga, K.A. Nguyen, S.J. Su, T.L. Windus, M. Dupuis and J.A. Montgomery, *J. Comp. Chem.* **14**, 1347 (1993).

[47] R. Cammi, B. Mennucci and J. Tomasi, *J. Am. Chem. Soc.* **34**, 8834 (1998).

[48] (a) S.R. Marder, J.W. Perry, B.G. Temann, C.B. Gorman, S. Gilmour, S.L. Biddle and G. Bourhill, *J. Am. Chem. Soc.*, **115**, 2524 (1993); (b) F. Meyers, S.R. Marder, B.M. Pierce and J.L. Bredas, *J. Am. Chem. Soc.* **116**, 10703 (1994); (c) S.R. Marder, C.B. Gorman, F. Meyers, J.W. Perry, G. Bourhill, J.L. Bredas and B.M. Pierce, *Science* **265**, 632 (1991).

[49] T.H. Dunning and P.J. Hay, in *Modern Theoretical Chemistry*, H.F. Schaefer III (ed.) (Plenum: New York, 1976); pp. 1-28.

[50] C. Reichardt, *Solvents and Solvent Effects in Organic Chemistry*, (VCH, Weinheim, 1990); second edition.

Part III

Relativistic models

Chapter 10

Variational methods in relativistic quantum mechanics: new approach to the computation of Dirac eigenvalues

Jean Dolbeault, Maria J. Esteban and Eric Séré
CEREMADE (UMR CNRS 7534),
Université Paris-Dauphine,
Place Maréchal Lattre de Tassigny,
F-75775 Paris Cedex 16, France.
{dolbeaul,esteban,sere}@ceremade.dauphine.fr

Abstract : The main goal of this paper is to describe some new variational methods for the characterization and computation of the eigenvalues and the eigenstates of Dirac operators. Our methods are all based on exact variational principles, both of min-max and of minimization types. The minimization procedure that we introduce is done in a particular set of functions satisfying a nonlinear constraint. Finally, we present several numerical methods that we have implemented in particular cases, in order to construct approximate solutions of that minimization problem.

1   Introduction

The free Dirac operator has been successfully used in the description of the kinematics of the electron. It is a first order operator which, in the appropriate units, has the form

$$H_0 = i\,\vec{\alpha}\cdot\vec{\nabla} + \beta \tag{1}$$

where $\vec{\alpha}\cdot\vec{\nabla} = \sum_{k=1}^{3} \alpha_k \partial_k$ and α_k are the Pauli-Dirac matrices,

$$\beta = \begin{pmatrix} I & 0 \\ 0 & -I \end{pmatrix}, \quad \alpha_k = \begin{pmatrix} 0 & \sigma_k \\ \sigma_k & 0 \end{pmatrix} \quad (k = 1,2,3),$$

σ_k being the well-known 2×2 matrices Pauli matrices :

$$\sigma_1 = \begin{pmatrix} 0 & 1 \\ 1 & 0 \end{pmatrix}, \quad \sigma_2 = \begin{pmatrix} 0 & -i \\ i & 0 \end{pmatrix}, \quad \sigma_3 = \begin{pmatrix} 1 & 0 \\ 0 & -1 \end{pmatrix}.$$

The spectrum of the free Dirac operator is only continuous and we have :

$$\sigma(H_0) = (-\infty, -1] \cup [1, +\infty).$$

The total unboundedness of $\sigma(H_0)$ creates many difficulties which are not present in its nonrelativistic limit, the Schrödinger operator, which is semibounded. The difficulties associated with the numerical computation of Dirac eigenvalues are known under the generic name of *variational collapse* .

The basic equation describing an electron evolving in an exterior scalar potential V is

$$i\,\partial_t\,\Psi = (H_0 + V)\Psi \quad \text{in} \quad \mathbb{R} \times \mathbb{R}^3. \tag{2}$$

When looking for stationary states of (2) of the form $\Psi(t, x) = e^{-i\lambda t}\,\psi(x)$, one checks that the wave function ψ has to satisfy the following stationnary equation:

$$(H_0 + V)\,\psi = \lambda\,\psi \quad \text{in} \quad \mathbb{R}^3. \tag{3}$$

Equality (3) is an eigenvalue equation which corresponds to bound states of the electron if ψ is square integrable in $\mathbb{R}^3$. For those states to be stable, the eigenvalue λ has to be in the gap of the spectrum of H_0, *i.e.* in $(-1, 1)$. In particular, if it exists, the smallest eigenvalue (or the smallest positive eigenvalue) of $H_0 + V$ in the gap $(-1, 1)$, corresponds to the ground-state level of the electron in the potential V (in the sense that its nonrelativistic limit is the ground-state level of the limiting Schrödinger equation).

In the case of a semibounded operator H like the Schrödinger operator, finding the ground-state corresponds to minimizing the Rayleigh quotient

$$R(\psi) := \frac{((H + V)\psi, \psi)}{(\psi, \psi)}, \tag{4}$$

where by $(\cdot, \cdot)$ we denote either the $L^2(\mathbb{R}^3)$ - inner product or, when necessary, a duality product. There are many practical ways of tackling the problem of minimizing a function like $R(\psi)$.

In the case of the Dirac potential H_0, and for standard potentials V, minimizing $R(\psi)$ is useless, because it takes us to $-\infty$. On the other hand, the solutions of (3) correspond to critical points of $R(\psi)$. Hence, one has to find ways of characterizing non-minimization variational problems to find the eigenvalues of $H_0 + V$ in the interval $(-1, 1)$. Many works have been devoted to this question. W. Kutzelnigg has written two excellent reviews on this subject, where many relevant references can be found : [15, 16]. The main approaches to this problem can be classified in three groups:

1) Use of approximate Hamiltonians: the first idea is to replace (3) by another equation in which the main operator is semibounded. In general this is done by reducing (3) to a system of two equations for the upper and lower spinors in ψ. Then, one eliminates the lower spinor and shows that this system is equivalent to an equation for the upper spinor which involves a semibounded Hamiltonian. For instance, the above reduction can be done via the Foldy-Wouthuysen transformation. Finally, approximate Hamiltonians are constructed by considering different expansions of the exact Hamiltonian in powers of $1/c^2$ or other small quantities. This method yields models which are perturbations of the nonrelativistic one. To this category of works belong for instance [8, 9, 17, 18, 16].

2) Using appropriate finite "basis", with the right asymptotic (or other) behavior and thus avoiding to fall into the negative continuum $(-\infty, -1)$. This is equivalent to projecting the equation onto a well chosen space. For instance, this has been done in [7, 15].

3) Finding variational approaches other than simple minimization like minimization of Rayleigh quotients for the squared Hamiltonian $(H_0 + V)^2$ (see [21, 1]) or later on, maximization of the Rayleigh quotients for the "inverse Hamiltonian" $((H_0 + V)\psi, \psi)/((H_0 + V)^2\psi, \psi)$ (see [13]).

Actually, it is rather easy to see that the eingenstates of $(H_0 + V)$ should be obtained by defining appropriate min-max procedures for the Rayleigh quotient. This idea was first put forward by Talman [19]. Then other proposals followed : [10, 11, 4, 12, 5]. In these papers, various min-max and constrained minimization procedures were described and justified for some families of potentials.

Our work places itself in the third direction, but still uses the idea of eliminating the lower spinor in terms of the upper one, without any further approximation. In this paper, we present three different alternatives for the characterization and computation of all the eigenvalues of $H_0 + V$ in the gap of the essential spectrum of $H_0 + V$. Of course, in order to define these variational procedures, one has to choose a class of potentials V for which the operator $H_0 + V$ is well defined. In some sense, V has to be not "too strong" with respect to H_0. The class of potentials for which our methods work include sums of powers $-\gamma_i |x|^{-\beta_i}$, β_i being positive, but satisfying $\beta_i \leq 1$. Hence, sums of Coulomb potentials are admissible. Even if it is not always necessary, our presentation will be simplified if we make the assumption that V is an attractive potential, *i.e.* V is nonpositive everywhere in $\mathbb{R}^3$. For more details about the precise assumptions which are necessary on V see below and [4, 5].

To end the introduction, let us now give an idea of what is the meaning of min-max variational arguments. Note that for a given smooth enough function f, its minimum points are places in which the derivative of f is equal to 0. Min-max points are other places in which the derivative is equal to 0, but around which the function f looks like a saddle point. For instance, imagine a mountain-pass which is a point at which a road crosses a mountain range when going from a valley to another one. The mountain-pass is a maximum point for the road, but in the direction of the crest of the mountain range, it is a minimum point. A mountain-pass is a typical example of min-max arguments. A not too technical reference for understanding the different kinds of variational methods is [14].

2   Min-max approaches.

As far as the Dirac operator is concerned, the first works in this direction are those of Talman [19] and Datta-Deviah [3]. In [19] Talman proposed the following strategy: if any 4-spinor $\psi \in \mathbb{C}^4$ is seen as a pair $\psi = \binom{\varphi}{\chi}$, with φ, χ taking values in $\mathbb{C}^2$, let us compute the following min-max

$$\inf_{\varphi \neq 0} \sup_{\chi} \frac{(\psi, (H_0 + V)\psi)}{(\psi, \psi)} . \tag{5}$$

Talman claimed that the above min-max yields in fact the ground-state energy of the operator $H_0 + V$. This assertion is very interesting, since the decomposition of any spinor ψ into its upper and lower components, φ and χ, is indeed much easier to implement than projections based on spectral decomposition of H_0. The dificulty here is to give appropriate conditions on the potential for the approach to be valid.

In [10] another min-max strategy was proposed. This was the first rigorous approach to the variational solution of the problem under study and we present an improved version of it below. Then, the first abstract min-max characterization for the eigenvalues of operators with gaps was given by Griesemer and Siedentop in [11]. In [11], we also find the first mathematical justification of the correctness of Talman's min-max for a class of bounded potentials V. Other results in this direction have been proved in [4, 12].

The min-max approach which seems to apply to the largest class of potentials is described in [5]. In that paper, we use an abstract variational method together with an appropriate continuation argument which allows us to treat all potentials V which are not too singular at the origin and which, being self-adjoint, have a smallest eigenvalue in the interval $(-1, 1)$ which "comes from the positive continuum", that is, which is close to 1 for small values of the coupling constant γ. In particular, this method is applicable to the Coulomb potentials $-\gamma |x|^{-1}$, in the optimal class $\gamma \in (0, 1)$.

The main result concerning the min-max in [5] is the following. Let us consider an operator $H_0 + \gamma V_1$, where V_1 is a given scalar potential and $\gamma > 0$ is a coupling constant. Assume that there is an orthogonal decomposition of $\mathcal{H} = L^2(\mathbb{R}^3, \mathbb{C}^4)$ as $\mathcal{H} = \mathcal{H}_+ \oplus \mathcal{H}_-$ and let us denote by $\Lambda_\pm$ the projectors associated to this decomposition.

Assume moreover that

(i)   there exists a dense subspace F of $H^1(\mathbb{R}^3, \mathbb{C}^4)$, such that $F_\pm := \Lambda_\pm F$ are two subspaces of $H^{1/2}(\mathbb{R}^3, \mathbb{C}^4)$.

(ii) $a := \displaystyle\sup_{\psi \in F_- \setminus \{0\}} \frac{(\psi, H_0 \psi)}{\|\psi\|_{\mathcal{H}}^2} < +\infty$.

Then, we define the min-max levels :

$$\lambda_{k,\gamma}(V_1) := \inf_{\substack{W \text{ subspace of } F_+ \\ \dim W = k}} \quad \sup_{\psi \in (W \oplus F_-) \setminus \{0\}} \frac{(\psi, (H_0 + \gamma V_1) \psi)}{\|\psi\|_{\mathcal{H}}^2} \,, \quad k \geq 1, \quad (6)$$

and we assume that

(iii) $\lambda_{1,0}(V_1) > a$.

On V_1 we make the hypothesis:

(iv) $V_1(x) \xrightarrow[|x| \to +\infty]{} 0$,

Then, if we define $b := \inf \{ \sigma_{ess}(H_0 + \gamma V_1) \cap [a, +\infty) \}$, we can state the following result (see [5]) :

Under the above assumptions (i)-(iv), for all $\gamma > 0$ such that $H_0 + \gamma V_1$ can be defined as a self-adjoint operator with domain included in $H^{1/2}(\mathbb{R}^3, \mathbb{C}^4)$ and such that $\lambda_{1,\gamma}(V_1) > a + \gamma \sup_{\mathbb{R}^3} V_1$, all the eigenvalues of $H_0 + \gamma V_1$ in the interval (a, b) are given by the sequence $\lambda_{k,\gamma}(V_1)$. In particular, the energy of the ground-state is equal to $\lambda_{1,\gamma}(V_1)$. Note that if there is no eigenvalue of $H_0 + \gamma V_1$ in the interval (a, b) then all the min-max values $\lambda_{k,\gamma}$ are equal to b.

So, in this result it is not necessary to assume that the potentials are attractive. In this particular case, the assumption $\lambda_{1,\gamma}(V_1) > a + \gamma \sup_{\mathbb{R}^3} V_1$ is of course replaced by $\lambda_{1,\gamma}(V_1) > a$.

Let us notice that a simple case in which are satisfied all the assumptions for the above result to hold is given by γ and V_1 satisfying:

$$-\frac{\nu}{|x|} - c_1 \leq V_1 \leq c_2 = \sup_{\mathbb{R}^3} V_1 < +\infty,$$

$$c_1, c_2 \geq 0, \quad \gamma(c_1 + c_2) < 1 + \sqrt{1 - \gamma^2 \nu^2},$$

and $a = -1$.

The above result can be particularized to various cases in which different decompositions of $\mathcal{H}$ are considered. A possible decomposition corresponds to the projectors associated with the free Dirac operator, that is, $\Lambda_+ = \chi_{(0,+\infty)}(H_0)$ and $\Lambda_- = \chi_{(-\infty,0)}(H_0)$. Another decomposition which seems simpler for actual computations corresponds to considering the upper and the lower spinors separately, i. e.:

$$\text{for } \psi = \begin{pmatrix} \varphi \\ \chi \end{pmatrix}, \quad \Lambda_+\psi = \begin{pmatrix} \varphi \\ 0 \end{pmatrix}, \quad \Lambda_-\psi = \begin{pmatrix} 0 \\ \chi \end{pmatrix}. \tag{7}$$

Let us also notice that with the above decomposition (7), $\lambda_{1,\gamma}(V_1)$ is the min-max proposed by Talman. In particular, the above result proves that Talman's min-max yields indeed the first eigenvalue of $H_0 + \gamma V_1$ for all potentials V_1 and all constants γ such that the above properties are satisfied. In the particular case of the Coulomb potentials $-\gamma|x|^{-1}$, this means being able to consider any γ between 0 and 1, the optimal range of self-adjointness. Note also that less singular potentials like $\gamma|x|^{-\beta}$ can be dealt with as long as $\beta \in (0,1)$ and γ is such that $\lambda_{1,\gamma}(V_1) > -1$.

3   Minimization method and corresponding min-max approaches.

3.1   A constrained minimization method.

In [4] we reduced the computation of the first eigenvalue of a large family of Dirac operators to the consideration of a minimization problem in a class of functions described by a nonlinear constraint. Below we describe this method, since we believe that it can be useful in numerical computations or at least in understanding the positivity properties of constrained problems.

First, one reduces the eigenvalue for the 4-spinor $\psi = \begin{pmatrix} \varphi \\ \chi \end{pmatrix}$ to a system of two equations for the 2-spinors φ and χ. This is a common procedure when trying to construct approximate semibounded Hamiltonians, as was described in the Introduction.

Let us introduce a shift of length -1 in the eigenvalues: $E = \lambda - 1$, so that for λ to be in the spectral gap $(-1,1)$, E has to be in the interval $(-2,0)$. Next, notice that the equation $(H_0 + V)\psi = \lambda\psi = (E+1)\lambda$ is equivalent to the system

$$\begin{cases} L\chi = (E - V)\varphi \\ L\varphi = (E + 2 - V)\chi \end{cases} \tag{8}$$

with $L = i(\vec{\sigma} \cdot \vec{\nabla}) = \sum_{k=1}^{3} i\sigma_k \partial/\partial x_k$. As long as $E + 2 - V \neq 0$, the system (8) can be written as

$$L\left(\frac{L\varphi}{g_E}\right) + V\varphi = E\varphi, \quad \chi = \frac{L\varphi}{g_E} \tag{9}$$

with $g_E = E + 2 - V$. Note that for attractive potentials V ($V \leq 0$ a. e.), $g_E \neq 0$ for all E in the gap $(-2,0)$.

Then, we consider $\phi = \varphi/\sqrt{g_E}$ which solves the equation

$$H_E \phi := \sqrt{g_E} \, L\Big(\frac{1}{g_E} L(\sqrt{g_E} \, \phi)\Big) = (E - V)(E + 2 - V)\phi \qquad (10)$$

where the operator H_E defined in (10) is symmetric.

Thus, any $E \in (-2,0)$, eigenvalue of the operator $H_0 + V - 1$, with associated eigenfunction $\varphi = \sqrt{g_E} \, \phi$, is a solution of the equation

$$(\phi, \phi) \, E^2 + 2(\phi, (1 - V)\phi)E - (\phi, (2 - V)V\phi) - (\phi, H_E \phi) = 0 \,, \qquad (11)$$

which is quadratic in E if we forget the dependence of H_E on E. So, E is necessarily solution of one of the following equations :

$$E = J^\pm(E, \phi) := \frac{1}{(\phi, \phi)}\Big(\pm \sqrt{\Delta(E, \phi)} - \big(\phi, (1 - V)\phi\big)\Big) \qquad (12)$$

where $\Delta(E, \phi) := |(\phi, V\phi)|^2 + (\phi, \phi)[(\phi, \phi) + (\phi, H_E \phi) - (\phi, V^2 \phi)]$.

Note that if $T(E, \phi) := [(\phi, \phi)][(\phi, \phi) + (\phi, H_E \phi) - (\phi, V^2 \phi)]$ is nonnegative, then the range of J^- (resp. J^+) is contained in the interval $(-\infty, -1]$ (resp. $[-1, +\infty)$). Hence, eigenvalues E corresponding to positive energies λ necessarily satisfy the equation $E = J^+(E, \phi)$.

By a simple continuation argument, one can see that for a large family of potentials, the inequality $T(E, \phi) \geq 0$ is equivalent to the existence of a gap around 0 in the spectrum of $H_0 + V$. Hence, one can try to compute the smallest positive eigenvalue of $H_0 + V$ by just minimizing the functional $J^+(E, \phi)$ in the set $\{(E, \phi) \, ; \, J^+(E, \phi) = E\}$. And indeed, we prove in [4] that under appropriate assumptions, the solution of this minimization problem is a ground-state for $H_0 + V - 1$ (at this level, we do not discuss the regularity conditions required on the functions ϕ). The class of potentials $-\gamma|x|^{-\beta}$ is again admissible for all $\beta \in (0, 1]$ and $\gamma > 0$ not too large. In the case $\beta = 1$, the condition on γ is again optimal : $0 < \gamma < 1$.

3.2   Relationship with Talman's min-max.

In Section 3, we introduced a general class of min-max problems yielding the eigenvalues of $H_0 + V$ in the interval $(-1, 1)$. Here, the same has been achieved for the positive ground state of the electron moving in the potential V. Actually, it is not difficult to see that there is a strong relationship between the two methods, specially in the case in which the decomposition of the 4-spinors is the one which corresponds to the consideration of the upper and lower 2-spinors separately. As pointed out above, this was actually the proposal made by Talman. Let us come back to that min-max for potentials $\gamma V_1, \gamma > 0$:

$$\lambda_1^\gamma = \inf_{\varphi \neq 0} \sup_\chi \frac{(\psi, (H_0 + \gamma V_1)\psi)}{(\psi, \psi)} \quad , \quad \psi = \begin{pmatrix} \varphi \\ \chi \end{pmatrix}. \tag{13}$$

If one tries to apply the result of Section 3, one sees that a can be chosen to be equal to -1. Moreover, one can explicitly solve the maximization problem in χ as follows : for every φ, the supremum

$$\lambda^\gamma(\varphi) := \sup_\chi \frac{(\psi, (H_0 + \gamma V_1)\psi)}{(\psi, \psi)} \quad , \quad \psi = \begin{pmatrix} \varphi \\ \chi \end{pmatrix} \tag{14}$$

is achieved by

$$\chi^\gamma(\varphi) := \frac{L\varphi}{1 - \gamma V_1 + \lambda^\gamma(\varphi)}, \tag{15}$$

and $\lambda^\gamma(\varphi)$ is the unique number in $(-1, +\infty)$ such that

$$\lambda^\gamma(\varphi) \int_{\mathbb{R}^3} |\varphi|^2 dx = \int_{\mathbb{R}^3} \left(\frac{|L\varphi|^2}{1 - \gamma V_1 + \lambda^\gamma(\varphi)} + (1 + \gamma V_1)|\varphi|^2 \right) dx, \tag{16}$$

So, finally, λ_1^γ can be defined as the minimum (in φ) of all $\lambda^\gamma(\varphi)$. This is again a minimization problem with a nonlinear constraint.

This minimization method is actually equivalent to the one described in Section 4.1 and they are both rigorously justified for the same class of potentials.

4   Some related numerical computations.

In this section we present recent numerical results that have been obtained for the very particular class of potentials $-\gamma|x|^{-\beta}$, with $\gamma > 0$, $0 < \beta \leq 1$ and in some specific cases (see [6]). Our aim has been to show that these new variational techniques could help to better understand how to perform computations in relativistic quantum mechanics, without having to care about variational collapse, boundary conditions, choice of good special basis of functions, etc.

More precisely, we have implemented three methods, one based on shooting arguments and the other two, on variational ones. The first one is used as a comparison test for the variational method. Note that for $\beta = 1$, the exact eigenvalues are explicitly known, wich provides us with a good test for the computations.

Let V be a radial attractive scalar potential and let us try compute the first positive eigenvalue of $H_0 + V$. Since V is radial (see for instance [20, 2]) the eigenfunctions of $H_0 + V$ can be expressed in terms of the spherical harmonics according to the decomposition

$$L^2(\mathbb{R}^3; \mathbb{C}^4) = L^2(]0, +\infty[, r^2 dr; \mathbb{C})$$
$$\bigotimes \left(\bigoplus_{j = \frac{1}{2}, \frac{3}{2}, \dots}^{+\infty} \bigoplus_{m_j = -j}^{+j} \bigoplus_{\kappa_j = \pm(j + \frac{1}{2})} \mathcal{K}_{m_j, \kappa_j} \right).$$

of the set of the square integrable functions defined on $\mathbb{R}^3$ with values in $\mathbb{C}^4$. Any spinor $\psi \in L^2(\mathbb{R}^3 \, \mathbb{C}^4)$ can therefore be written as

$$\psi(x) = \sum_{\substack{j, m_j, \kappa_j \\ \epsilon = \pm}} \frac{1}{|x|} f^\epsilon_{m_j, \kappa_j}(|x|) \Phi^\epsilon_{m_j, \kappa_j}\left(\frac{x}{|x|}\right) \tag{17}$$

where

$$\Phi^+_{m_j, \mp(j+1/2)} = \begin{pmatrix} i\Psi^{m_j}_{j\mp1/2} \\ 0 \end{pmatrix}, \quad \Phi^-_{m_j, \mp(j+1/2)} = \begin{pmatrix} 0 \\ \Psi^{m_j}_{j\pm1/2} \end{pmatrix} \tag{18}$$

generate the space $\mathcal{K}_{m_j, \kappa_j}$ and can be expressed in terms of the spherical harmonics as follows :

$$\Psi^{m_j}_{j-1/2} = \frac{1}{\sqrt{2j}} \begin{pmatrix} \sqrt{j + m_j} \, Y^{m_j-1/2}_{j-1/2} \\ \sqrt{j - m_j} \, Y^{m_j+1/2}_{j-1/2} \end{pmatrix}, \tag{19}$$

$$\Psi^{m_j}_{j+1/2} = \frac{1}{\sqrt{2j+2}} \begin{pmatrix} \sqrt{j+1-m_j} \, Y^{m_j-1/2}_{j+1/2} \\ -\sqrt{j+1+m_j} \, Y^{m_j+1/2}_{j+1/2} \end{pmatrix}, \tag{20}$$

where $(Y^m_l)^{m=-l,-l+1,...l}_{l=0,1,2...}$ are the usual spherical harmonics. The radial Dirac operator acting on the set of the square integrable real functions on $(0, +\infty)$, $L^2(0, +\infty)$, is

$$h_\kappa = \begin{pmatrix} 1+V & -\dfrac{d}{dr} + \dfrac{\kappa}{r} \\ \dfrac{d}{dr} + \dfrac{\kappa}{r} & -1+V \end{pmatrix} \quad \kappa = \pm1, \pm2, ... \tag{21}$$

and the eigenvalue problem takes the form

$$\begin{cases} u' = (1+\lambda)v - (Vv + \dfrac{\kappa}{r}u) \\ v' = (1-\lambda)u + (Vu + \dfrac{\kappa}{r}v) \end{cases} \tag{22}$$

The solutions of this system are characterized by two parameters, λ and $\delta = v(1)/u(1)$ for instance, and we shall denote by X the set of the solutions of (22) such that $u(1) = 1$ when λ and δ vary in $\mathbb{R}$. However, the condition that u and v are in $L^2(0, +\infty)$ determines uniquely λ and δ. One can show that this integrability condition is equivalent to assuming that

$$\lim_{r \to 0_+} r(|u(r)|^2 + |v(r)|^2) = 0,$$
$$\lim_{r \to +\infty}(|u(r)|^2 + |v(r)|^2) = 0, \tag{23}$$

thus providing a simple numerical ("shooting") method to determine λ and δ (we shall refer to this method by the letter "s" and use it to compare the numerical results with the numerical minimization method given below).

Let us describe now a first numerical minimization method, which uses the special form (22) of the eigenvalue problem and goes along the main lines of the method described in Section 4.1. Similarly to (9), in (22) v can be eliminated in terms of u:

$$\frac{v}{r^\kappa} = (r^{2\kappa}(1 + \lambda - V))^{-1}\frac{d}{dr}(r^\kappa u). \tag{24}$$

For many central potentials, the ground-state of $H_0 + V$ is a solution of (22), with $\kappa = -1$. For instance, in the case of Coulomb potentials, there is no square integrable solution of (22) when $\kappa = 1$ and the ground-state is achieved for $\kappa = -1$. The eigenvalue problem (22) for $\kappa = -1$ is now equivalent to solving

$$h^\lambda \phi = (1 + \lambda - V)(1 - \lambda + V)\phi \tag{25}$$

where h^λ is a symmetric operator

$$h^\lambda \phi = \sqrt{1 + \lambda - V} \, \frac{d}{dr} \left[\frac{r^2}{1 + \lambda - V} \frac{d}{dr} (\sqrt{1 + \lambda - V} \phi) \right] \tag{26}$$

and $\phi(r) = r^{-1} u(r)/\sqrt{1 + \lambda - V}$ is now a function defined on $(0, +\infty)$. Equation (10) is then equivalent to

$$(\phi, \phi)\lambda^2 - 2(\phi, V\phi)\lambda + (\phi, V^2\phi) - [(\phi, \phi) + (\phi, h_\lambda \phi)] = 0 \tag{27}$$

where $(\cdot, \cdot)$ is the usual scalar product in $L^2(0, +\infty)$ (and $\| \cdot \|$ the corresponding norm). The eigenvalue problem is then reduced to finding a critical point of $J^+(\lambda - 1, \cdot)$ with $\lambda - 1 = J^+(\lambda - 1, \phi)$ and

$$J^+(\lambda - 1, \phi) + 1 = \frac{\sqrt{\Delta(\lambda - 1, \phi)} + (\phi, V\phi)}{\|\phi\|^2}, \tag{28}$$

$$\Delta(\lambda - 1, \phi) = (\phi, V\phi)^2 + \|\phi\|^2 [(\phi, h_\lambda \phi) + \|\phi\|^2 - (\phi, V^2\phi)]. \tag{29}$$

To solve this constrained problem numerically, the natural idea is to introduce a penalization method and to minimize

$$J^+(\lambda - 1, \phi) + A|(\lambda - 1) - J^+(\lambda - 1, \phi)|^2$$

in the limit $A \to +\infty$. In practical computations, one has to consider positive constants A large enough to ensure that the constraint is "almost" satisfied. Actually if we assume that ϕ is given by (26) with (u, v) in X, the condition that u and v are in $L^2(0, +\infty)$ is equivalent to assuming that the integrals involved in the expression (28) are finite. Of course these integrals are numerically computed on an interval (ϵ, R), because there is a singularity at $r = 0$ and one wants to compute the integral in a finite interval. So, the approximate value $J_{\epsilon,R}^+$ of J^+ is finite even if the constraint is not satisfied, but we observe that $\lim_{(\epsilon, R) \to (0, +\infty)} J_{\epsilon,R}^+(\lambda - 1, \phi) = +\infty$ unless $\lambda - 1 = J^+(\lambda - 1, \phi)$. Hence, a minimization of J^+ (numerically of $J_{\epsilon,R}^+$) on the set X takes care of the constraint $\lambda - 1 = J^+(\lambda - 1, \phi)$ automatically. This method will be referred by the letter "m" in Tables 10.1 and 10.2. Note that from a mathematical point of view, this is also a shooting method in (λ, δ).

In Table 1 below we present a comparison of the shooting (s) and the minimization (m) methods for $\kappa = -1$, $V(r) = -\gamma r^{-\beta}$, $\gamma = 0.5$ and $\beta \in (0, 1)$. The system (22) is numerically solved with a stepsize adaptative Runge-Kutta method on the interval $(\epsilon = 10^{-4}, R = 15)$. For the shooting method we minimize the quantity $\epsilon(|u(\epsilon)|^2 + |v(\epsilon)|^2) + \theta(|u(R)|^2 + |v(R)|^2) = \Delta_s$ for some scale parameter

Table 10.1:

β	δ_s	δ_m	λ_s	λ_m	J^+	CEr	Δ_s
1	-0.267954	-0.267943	0.866034	0.866013	0.866014	$1.8\ 10^{-12}$	0.00029
0.9	-0.235187	-0.235174	0.856725	0.856698	0.856698	$2.1\ 10^{-14}$	0.00053
0.8	-0.207802	-0.207788	0.843181	0.843146	0.843146	$5.2\ 10^{-14}$	0.00063
0.7	-0.183397	-0.183379	0.825877	0.825832	0.825831	$4.3\ 10^{-13}$	0.00076
0.6	-0.160651	-0.160627	0.804699	0.804639	0.804639	$4.1\ 10^{-13}$	0.00094
0.5	-0.138654	-0.138619	0.779161	0.779071	0.779070	$3.4\ 10^{-13}$	0.0012
0.4	-0.116645	-0.116584	0.748381	0.748221	0.748220	$3.8\ 10^{-13}$	0.0018
0.3	-0.0938375	-0.0937016	0.710904	0.710537	0.710536	$3.5\ 10^{-13}$	0.0049
0.2	-0.069224	-0.068798	0.664252	0.663067	0.663067	$2.4\ 10^{-13}$	0.0097
0.1	-0.0412322	-0.0392963	0.60391	0.59833	0.59833	$1.4\ 10^{-13}$	0.018

$\theta > 0$, while for the minimization method, the quantity $J^+(\lambda - 1, \phi)$ is directly minimized, the quantity $CEr := |J^+(\lambda - 1, \phi) - (\lambda - 1)|^2$ being computed a posteriori. The parameter θ is chosen in order that the terms $\epsilon(|u(\epsilon)|^2 + |v(\epsilon)|^2)$ and $\theta(|u(R)|^2 + |v(R)|^2)$ have the same maximum value on the boundary of the domain of minimization. For $\beta = 1$, the result is known explicitly: $\lambda_1 = [1 - \gamma^2]^{1/2} = 0.866025...$, $\delta_1 = -[(1 - \lambda)/(1 + \lambda)]^{1/2} = -0.267949....$ For practical reasons, the results given here correspond to parameters taken in a neighborhood of (δ_1, λ_1). The results correspond therefore to the branch $(\delta_\beta, \lambda_\beta)$ starting from (δ_1, λ_1) at $\beta = 1$ and parametrized by β.

The main advantage of the minimizing approach is that it can be extended to the case of nonsymmetric (non central) potentials, but of course for a minimizing set which is larger than X.

We will now assume that the potential is radial, but consider a general basis of $L^2(0, +\infty)$ (of course well chosen). For that purpose, we introduce a third formulation, which is intermediate between the abstract min-max theory and the minimization of J^+, and goes as follows. Its main advantage is that the constraint $E = J^+(E, \phi)$ will then be automatically satisfied. We will therefore call this method the "direct minimisation method".

As in (9), we may rewrite (8) as

$$L\left(\frac{L\varphi}{\lambda + 1 - V}\right) + V\varphi = (\lambda - 1)\varphi, \quad \chi = \frac{L\varphi}{\lambda + 1 - V}, \tag{30}$$

at least if $\lambda \in (-1, 1)$ and if V is nonpositive almost everywhere. Multiplying (30) by φ and integrating with respect to $x \in \mathbb{R}^3$, we obtain :

$$f_\varphi(\lambda) := \int_{\mathbb{R}^3} \frac{|L\varphi|^2}{\lambda + mc^2 - V}\, dx \tag{31}$$

$$= \quad (\lambda - 1) \int_{\mathbb{R}^3} |\varphi|^2 \, dx - \int_{\mathbb{R}^3} V |\varphi|^2 \, dx =: g_\varphi(\lambda) \, .$$

Since for a given φ, $f_\varphi(\lambda)$ is decreasing and $g_\varphi(\lambda)$ is increasing in λ, if there exits a $\lambda = \lambda(\varphi)$ such that (31) is satisfied, then it is unique (the existence of such a λ for all φ depends on the properties of the potential V). According to Section 4.2, for those V's, the ground state is such that

$$\lambda_1 = \min_\varphi \lambda(\varphi) \, . \tag{32}$$

One can solve (32) by any numerical minimization method. A possible way to do it to consider a finite basis $\{\varphi_1, \ldots, \varphi_n\}$ and define $\lambda(x_1, \ldots, x_n)$ by :

$$\ell(x_1, \ldots, x_n) := \lambda \left(\sum_{i=1}^n x_i \, \varphi_i \right).$$

Then,

$$\lambda_1^n := \inf_{(x_1, \ldots, x_n) \in \mathbb{R}^n} \ell(x_1, \ldots, x_n)$$

is an approximation of $\lambda_1(V)$ which can be found by any well suited minimization algorithm.

In order ot simplify the presentation, we come back to the radially symmetric situation in which the potential V is central and we can decompose the whole problem by using spherical spinors. This is not necessary, but has the advantage of being easier to describe.

For a radial potential we may use the radial Dirac equation and consider (22) instead of (8). Define $\lambda = \lambda_r(u)$ as the unique solution of

$$f(\lambda) = \quad \int_0^{+\infty} \frac{|(r^\kappa u)'|^2}{r^{2\kappa}(1 + \lambda - V(r))} \, dr \tag{33}$$

$$= \quad (\lambda - 1) \int_0^{+\infty} |u(r)|^2 \, dr - \int_0^{+\infty} V(r)|u(r)|^2 \, dr := g(\lambda) \, .$$

(Notice that the existence of $\lambda_r(u)$ depends on the assumptions made on the potential V). Then, λ_1 is given by :

$$\lambda_1 = \inf_u \lambda_r(u) \, .$$

To solve (33) numerically it is more convenient to rewrite $f(\lambda)$ as an alternated series :

$$f(\lambda) = \sum_{k=0}^{+\infty} \left[(-1)^k \int_0^{+\infty} \frac{r^{-2\kappa}|(r^\kappa u)'|^2}{(1 - V(r))^{k+1}} \, dr \right] \lambda^k \, . \tag{34}$$

From a numerical point of view, the solution (with $\kappa = -1$) is approximated on a finite basis of $L^2(0, \infty)$, $(u_i)_{i=1,2,\ldots n}$: $u = \sum_{i=1}^n x_i u_i$. If

$$f_{ijk} = (-1)^{k-1} \int_0^{+\infty} \frac{r^2 (u_i/r)'(u_j/r)'}{(1 - V(r))^k} \, dr \tag{35}$$

Table 10.2:

β	0.90	0.93	0.95	0.97	0.99	1.00
$\lambda_1^{10,14}$	0.855681	0.858516	0.860228	0.861792	0.863200	0.863843
$\lambda_1^{10,15}$	0.858012	0.861112	0.863004	0.864749	0.866338	0.867071
J^+	0.856698	0.859984	0.861954	0.863735	0.865310	0.866014
Δ_{m}	0.0082	0.0058	0.0046	0.0033	0.0020	0.0022

and

$$V_{ij} = \int_0^{+\infty} u_i(r)u_j(r)V(r)\,dr \,, \tag{36}$$

the approximating equation for λ corresponding to (33) is then

$$\sum_{i,j=1}^{n} \left((\sum_{k=1}^{m} f_{ijk}\lambda^{k-1}) + V_{ij} \right) x_i x_j + (1-\lambda)\sum_{i=1}^{n} x_i^2 = 0 \,, \tag{37}$$

where the series in λ has been truncated at order m. It is actually more convenient to define

$$A^{n,m}(\lambda) = \left((\sum_{k=1}^{m} f_{ijk}\lambda^{k-1}) + (1-\lambda)\delta_{ij} + V_{ij} \right)_{i,j=1,2,\ldots n}$$

and to approximate λ_1 by $\lambda_1^{n,m}$ defined as the first positive root of $\lambda \mapsto \mu_1(\lambda) := \mu(A^{n,m}(\lambda))$ where $\mu(A)$ denotes the first eigenvalue of the matrix A. The function $\lambda \mapsto \mu_1(\lambda)$ is indeed continuous, nonincreasing in λ and such that $\mu_1(0) > 0$ when we make the right assumptions on the potential V, i.e. when V is not too strong with respect to H_0 (this corresponds to the hypothesis made in Section 3). Moreover, if there exist $x_1,\ldots,x_m$ such that (37) holds, then $\mu_1(\lambda) \le 0$. Hence, indeed $\lambda_1^{n,m} = inf\{\lambda > 0\,;\ \mu_1(\lambda) \le 0\}$.

Note that $(\lambda_1^{n,m} - \lim\limits_{m} \lambda_1^{n,m})_{m\ge1}$ is an alternating sequence (essentially converging at a geometric rate): consider indeed $u = \sum\limits_{i=1}^{n} x_i\,u_i$.

$$\sigma_m^n(\lambda) = \sum_{\substack{i,j=1,\ldots n \\ k=1,\ldots m}} x_i x_j f_{ijk}\lambda^{k-1} = \sum_{k=1}^{m} \int_0^{+\infty} (-1)^{k-1}\frac{r^2|(u/r)'|^2}{(1-V)^k}\lambda^{k-1}\,dr$$

$$= \int_0^{+\infty} \frac{r^2}{1-V}\,|(u/r)'|^2\,\frac{1 - \left(\dfrac{-\lambda}{1-V}\right)^m}{1 + \dfrac{\lambda}{1-V}}\,dr.$$

If $V \le 0$ a.e., $(1-V) \ge 1$, so that the series $(\sigma_m^n(\lambda))_{m\in\mathbb{N}}$ is an alternating sequence (and $(\lambda/(1-V))^m$ converges at a geometric rate).

The results in Table 2 have been obtained by taking an orthonormal basis generated by the ground state of the hydrogen atom and $n-1$ Hermite functions, with $n = 10$. Our purpose in this numerical computation was not to provide very accurate results but just to prove the feasibility of such a numerical approach. Clearly, depending on the specific properties of the potential, the choice of a well suited basis should greatly improve the accuracy of the computation.

More precisely, we have considered $V(r) = -\gamma r^{-\beta}$, $\gamma = 0.5$ and β close to 1. The approximating space is of dimension $n = 10$ and the series are truncated at $m = 14$ or $m = 15$ (the corresponding values $\lambda_1^{10,14}$ and $\lambda_1^{10,15}$ are respectively a lower and an upper bound of $\lim_{m\to+\infty} \lambda_1^{10,m}$). As in Table 1, J^+ is obtained through a minimization procedure on the set X, and Δ_m measures the error (in the L^2-norm) when the corresponding solution is approximated on the basis (with $n = 10$ elements) used for the direct minimisation method.

Bibliography

[1] W.E. Bayliss, S. J. Peel. Stable variational calculations with the Dirac Hamiltonian. Physical Review A, **28**(4) (1983), p. 2552-2554.

[2] J.D. Bjorken, S.D. Drell. *Relativistic quantum fields*. McGraw-Hill (1965).

[3] S.N. Datta, G. Deviah. The minimax technique in relativistic Hartree-Fock calculations. Pramana **30**(5) (1988), p. 393-416.

[4] J. Dolbeault, M.J. Esteban and E. Séré. Variational characterization for eigenvalues of Dirac operators. To appear in Calc. Var. and P.D.E.

[5] J. Dolbeault, M.J. Esteban, E. Séré. On the eigenvalues of operators with gaps. Application to Dirac operators. To appear in J. Funct. Anal.

[6] J. Dolbeault, M.J. Esteban, E. Séré, M. Vanbreugel. A minimization method for the one-particle Dirac equation. Submitted for publication.

[7] G.W.F. Drake and S.P. Goldman. Relativistic Sturmian and finite basis set methods in atomic physics. Adv. Atomic Molecular Phys., **25** (1988), p. 393-416.

[8] Ph. Durand. Transformation du Hamiltonien de Dirac en Hamiltoniens variationnels de type Pauli. Application à des atomes hydrogenoïdes. C. R. Acad. Sc. Paris **303**, série II, numéro 2 (1986), p. 119-124.

[9] Ph. Durand, J.-P. Malrieu. Effective Hamiltonians and pseudo potentials as tools for rigorous modelling. In *Ab initio methods in Quantum Chemistry I*. K.P. Lawley ed. J. Wiley and sons, 1987.

[10] M.J. Esteban, E. Séré. Existence and multiplicity of solutions for linear and nonlinear Dirac problems. *Partial Differential Equations and Their Applications*. CRM Proceedings and Lecture Notes, volume 12. Eds. P.C. Greiner, V. Ivrii, L.A. Seco and C. Sulem. AMS, 1997.

[11] M. Griesemer, H. Siedentop. A minimax principle for the eigenvalues in spectral gaps. J. London Math. Soc. **60**(2) (1999), p. 490-500.

[12] M. Griesemer, R.T. Lewis, H. Siedentop. A Minimax Principle for Eigenvalues in Spectral Gaps : Dirac Operators with Coulomb potentials. Doc. Math. **4** (1999), p. 275-283 (electronic).

[13] R.N. Hill, C. Krauthauser. A solution to the problem of variational collapse for the one-particle Dirac equation. Phys. Rev. Lett. **72**(14) (1994), p. 2151-2154.

[14] O. Kavian. *Introduction à la théorie des points-critiques.* Springer-Verlag, 1993

[15] W. Kutzelnigg. Basis Set Expansion of the Dirac Operator without Variational Collapse. Int. J. Quant. Chem. **25** (1984), p. 107-129.

[16] W. Kutzelnigg. Relativistic one-electron Hamiltonians for electrons only and the variational treatment of the Dirac equation. Chem. Phys. **225** (1997), p. 203-222.

[17] A. Le Yaouanc, L. Oliver, J.-C. Raynal. The Hamiltonian $(p^2 + m^2)^{1/2} - \alpha/r$ near the critical value $\alpha_c = 2/\pi$. J. Math. Phys. **38**(8) (1997), p. 3397-4012.

[18] E. van Lenthe, R. van Leeuwen, E.J. Baerends, J.G. Snijders. Relativistic regular two-component Hamiltonians. In *New challenges in computational Quantum Chemistry.* R. Broek et al ed. Publications Dept. Chem. Phys. and Material sciences. University of Groningen, 1994.

[19] J.D. Talman. Minimax principle for the Dirac equation. Phys. Rev. Lett. **57**(9) (1986), p. 1091-1094.

[20] B. Thaller. *The Dirac equation.* Springer-Verlag, 1992.

[21] H. Wallmeier, W. Kutzelnigg. Use of the squared Dirac operator in variational relativistic calculations. Chem. Phys. Lett. **78**(2) (1981), p. 341-346.

Chapter 11

Quaternion symmetry of the Dirac equation

T. Saue[1] **& H. J. Aa. Jensen**[2]
[1] *The Institute of Chemistry,*
University of Tromsø,
N-9037 Tromsø, Norway.
Trond.Saue@chem.uit.no
[2] *Department of Chemistry,*
University of Southern Denmark - Main campus: Odense University,
DK-5230 Odense M, Denmark.

Abstract : Following van der Waerden, the Dirac equation is derived from linearization of the Klein-Gordon equation using the algebraic properties of the Pauli spin matrices. As the algebra of these matrices is identical to that of quaternions, the Dirac equation can be reformulated in terms of quaternion algebra and therefore without reference to a specific spin quantization axis. In this paper we consider the symmetry content of the Dirac equation. It is found that the basic binary symmetry operations in spin space map onto the unit vectors of complex quaternions. We argue that a consistent choice of the inversion operator in spin space is of order four. We furthermore show that quaternion algebra is the natural language for time reversal symmetry. These considerations lead to the formulation of a symmetry scheme that automatically provides maximum point group and time reversal symmetry reduction in the solution of the Dirac equation in the finite basis approximation.

One day, while we were walking on the beach, he [Dirac] told me that he would teach me a saying:'It is easy, if you remember the symmetry. Watch the symmetry.' He went on, 'When a man says yes, he means perhaps; when he says perhaps, he means no; when he says no, he is no diplomat. When a lady says no, she means perhaps; when she says perhaps, she means yes; when she says yes, she is no lady'. With a couple of repetitions, I learned it, and he was pleased.
S.A. KURSUNOGLU (1987) [1]

1   Introduction

In 1878[2] and 1880[3], Frobenius and Peirce proved that the only associative real division algebras are real numbers, complex numbers, and real quaternions, that is they are the only numbers with a multiplicative inverse. A symmetry operation has always an inverse and so group representations are restricted to the above algebras. Quaternion algebra has, however, found very limited use in the physical sciences, perhaps due to the rise and predominance of vector analysis or the non-commutivity of quaternion multiplication. One may cite Kelvin[4]: *Quaternions came from Hamilton after his really good work had been done; and, though beautiful ingenious, have been an unmixed evil to those who have touched them in any way, including Clerk Maxwell.*

In non-relativistic quantum mechanics symmetry is generally handled by real algebra, whereas the spin-orbit coupling enforces the introduction of complex algebra in the relativistic domain. Yet the fundamental equation of relativistic quantum mechanics, the Dirac equation, has a manifestly quaternion structure. One should therefore expect that symmetry should rather be handled by quaternion algebra. In this paper we analyze the symmetry content of the Dirac equation and show that all three algebras – real, complex and quaternion – come into play in the relativistic domain. The actual choice of algebra is decided by the symmetry of the problem at hand.

2   The Dirac equation

The relativistic expression for the energy of a free particle is

$$E^2 = m^2c^4 + c^2p^2 \tag{1}$$

Straightforward operator substitution, that is quantization, leads to the Klein-Gordon equation

$$\left[\frac{1}{c^2} \frac{\partial^2}{\partial t^2} + \mathbf{p}^2 \right] \phi = -\left(mc \right)^2 \phi \tag{2}$$

The energy operator appears squared in this equation. Hence it has solutions of both positive and negative energies. Contrary to the classical case, the negative energy solutions cannot be discarded since the functional space would then become incomplete[5].

A troublesome feature of this equation is that the presence of the second derivative with respect to time makes $\phi^*\phi$ time dependent, and so it cannot be interpreted as a probability density. Dirac obtained a linearization of (2) by the introduction of the Dirac α and β matrices

$$\alpha = \begin{bmatrix} 0 & \sigma \\ \sigma & 0 \end{bmatrix}; \qquad \beta = \begin{bmatrix} I_2 & 0 \\ 0 & -I_2 \end{bmatrix}; \qquad [\alpha_q, \beta]_+ = 0, \quad q = x, y, z \qquad (3)$$

We shall derive the Dirac equation following an approach introduced by van der Waerden [6]. We expand the scalar wave function in (2) using Pauli matrices

$$\left[\frac{1}{c^2} \frac{\partial^2}{\partial t^2} + \mathbf{p}^2 \right] \phi = \left[\frac{i}{c} \frac{\partial}{\partial t} - (\sigma \cdot \mathbf{p}) \right] \left[-\frac{i}{c} \frac{\partial}{\partial t} - (\sigma \cdot \mathbf{p}) \right] \phi = -(mc)^2 \phi \qquad (4)$$

where ϕ is a two-component wave function. To obtain a first-order equation we introduce

$$\phi_1 = \frac{-1}{mc} \left[-\frac{i}{c} \frac{\partial}{\partial t} - (\sigma \cdot \mathbf{p}) \right] \phi; \qquad \phi_2 = \phi \qquad (5)$$

The second-order equation then becomes equivalent to two coupled first-order equations

$$\begin{aligned}
\left[\frac{i}{c} \frac{\partial}{\partial t} - (\sigma \cdot \mathbf{p}) \right] \phi_1 &= mc\phi_2 & (a) \\
\left[-\frac{i}{c} \frac{\partial}{\partial t} - (\sigma \cdot \mathbf{p}) \right] \phi_2 &= -mc\phi_1 & (b)
\end{aligned} \qquad (6)$$

The linearization of the Klein-Gordon equation must necessarily lead to coupled equations and not an eigenvalue problem involving a scalar function as in the non-relativistic domain. In the Schrödinger equation the momentum operator $\mathbf{p}$ appears squared and is therefore invariant under the operation of inversion; this is not possible by linearization of the relativistic energy expression. The functions ϕ_1 and ϕ_2 are not eigenfunctions of parity. Rather, the parity operator takes ϕ_1 into ϕ_2 and vice versa.

The two equations in (6) decouple for rest mass $m = 0$. (6b) was therefore proposed as the wave equation for a massless spin-$\frac{1}{2}$ particle in 1929 by Weyl [7], but was rejected since the wave function ϕ_2 is not invariant under parity. Parity is, however, not conserved in weak interactions. With the demonstration in 1957 of the violation of parity conservation in the β-decay of the ^{60}Co - nucleus [8], the Weyl equation was revived as a two-component equation for the neutrino.

To obtain the Dirac equation for the free electron we first take sums and differences

$$\frac{i}{c}\frac{\partial}{\partial t}\left[\phi_1 + \phi_2\right] - (\boldsymbol{\sigma} \cdot \mathbf{p})\left[\phi_1 - \phi_2\right] = mc\left[\phi_1 + \phi_2\right] \qquad (a - b)$$

$$(\boldsymbol{\sigma} \cdot \mathbf{p})\left[\phi_1 + \phi_2\right] - \frac{i}{c}\frac{\partial}{\partial t}\left[\phi_1 - \phi_2\right] = mc\left[\phi_1 - \phi_2\right] \qquad -(a + b)$$

$$(7)$$

and introduce the notation

$$\psi^L = \phi_1 + \phi_2; \quad \psi^S = \phi_1 - \phi_2 \tag{8}$$

We then obtain

$$\begin{bmatrix} \dfrac{i}{c}\dfrac{\partial}{\partial t} & -(\boldsymbol{\sigma} \cdot \mathbf{p}) \\[2ex] (\boldsymbol{\sigma} \cdot \mathbf{p}) & -\dfrac{i}{c}\dfrac{\partial}{\partial t} \end{bmatrix} \begin{bmatrix} \psi^L \\[1ex] \psi^S \end{bmatrix} = mc \begin{bmatrix} \psi^L \\[1ex] \psi^S \end{bmatrix} \tag{9}$$

The 4-component equation can be completely rewritten in terms of 4-vectors and scalar quantities as

$$(i\gamma_\mu \partial_\mu - mc)\,\psi = 0; \qquad \partial_\mu = \left(\nabla, -\frac{i}{c}\frac{\partial}{\partial t}\right), \quad \gamma_\mu = (\beta\boldsymbol{\alpha}, i\beta), \quad \psi = \begin{bmatrix} \psi^L \\ \psi^S \end{bmatrix} \tag{10}$$

and is therefore manifestly Lorentz invariant. The Dirac equation in its more familiar form is straightforwardly obtained by multiplication with βc from the left

$$\left[i\frac{\partial}{\partial t} - c\,(\boldsymbol{\alpha} \cdot \mathbf{p})\right]\psi = \beta mc^2 \psi \tag{11}$$

External fields are introduced by means of the substitutions

$$\mathbf{p} \to \boldsymbol{\pi} = \mathbf{p} + e\mathbf{A}; \quad E \to E + e\phi \tag{12}$$

where $\mathbf{A}$ and ϕ are vector and scalar potentials, respectively. The Dirac equation then attains the form

$$\hat{\mathcal{D}}\psi = \left[\hat{h}_D - i\frac{\partial}{\partial t}\right]\psi = 0; \qquad h_D = \beta mc^2 + c\,(\boldsymbol{\alpha} \cdot \boldsymbol{\pi}) - e\phi \tag{13}$$

3   The full symmetry group

The full *symmetry group* of a molecular system within the Born-Oppenheimer approximation consists of all symmetry operations $\hat{G}$ that commute with the electronic Hamiltonian of the system:

$$\left[\hat{G}, \hat{H}\right] = 0 \tag{14}$$

Symmetry operations are either unitary or antiunitary, as shown by the following argument by Wigner [9]: Observables calculated from a given wave function are invariant under any symmetry operation on the wave function. For the transition probability we must therefore have

$$\left|\left\langle \hat{G}\Psi_i \mid \hat{G}\Psi_j \right\rangle\right|^2 = \left|\langle \Psi_i \mid \Psi_j \rangle\right|^2 = \langle \Psi_i \mid \Psi_j \rangle \langle \Psi_i \mid \Psi_j \rangle^* \tag{15}$$

where the interaction operator $\hat{\Omega}$ has been set equal to one for simplicity (to obtain a totally symmetric operator). The above relation can be realized by

$$\begin{aligned}
\left\langle \hat{G}\Psi_i \mid \hat{G}\Psi_j \right\rangle &= \langle \Psi_i \mid \Psi_j \rangle; \quad \Rightarrow \quad \hat{G} \text{ is unitary} \\
\left\langle \hat{G}\Psi_i \mid \hat{G}\Psi_j \right\rangle &= \langle \Psi_j \mid \Psi_i \rangle; \quad \Rightarrow \quad \hat{G} \text{ is anti-unitary}
\end{aligned} \tag{16}$$

All spatial symmetry operations are unitary, and in molecular systems symmetry can be handled by use of the unitary representations of the molecular point groups. When antiunitary operations are included, the full symmetry cannot be represented with a set of representation matrices for each irreducible representation (irrep). However, Wigner has shown that we may still form a set of matrices, a *corepresentation*[9, 10, 11], which may be broken down to irreducible forms. We refer to our previous work [12] for a discussion of that approach.

3.1   Spatial symmetry

In this section we consider unitary symmetry operations, more specifically rotations, reflections and inversion of the space and spin coordinates. In relativistic theory spatial symmetry is usually handled by double group theory. Double groups were introduced as an artifice by Bethe [13] to avoid two-valued representations of fermion functions, which are not true representations of the symmetry group, and thereby recover the whole machinery of group theory. He introduced an extra element $\overline{E}$, corresponding to a rotation 2π about an arbitrary axis. This leads to a doubling of the number of symmetry operations of the group, but generally not to a doubling of the number of irreps. The extra irreps that appear in the double groups are spanned by fermion functions and are consequently denoted fermion irreps, whereas the irreps of the corresponding single groups are boson irreps.

We shall restrict the discussion to D_{2h} and subgroups. This is the set of all single point groups with no elements of order higher than two, and we shall therefore denote them *binary groups*. The binary groups are particularly simple to discuss since their single groups are all Abelian. Also, many computer codes limit point group symmetry to the binary groups using bit operations. The boson irreps of the binary groups are spanned by any scalar, the coordinates (x, y, z), the corresponding rotations (R_x, R_y, R_z) and the product of coordinates xyz. We shall denote the corresponding boson irreps Γ_0, Γ_q, Γ_{R_q}, and Γ_{xyz} $(q = x, y, z)$, respectively.

In terms of real scalar functions ("orbitals" in non-relativistic theory) any Dirac spinor has eight degrees of freedom, corresponding to the real and imaginary parts of the two large and the two small components. It is important to realise that although 4-spinors span fermion irreps, each of the eight real scalar functions of the spinor belongs to a specific boson irrep. Specifically, the symmetry content of the large components ψ^L is given by[12]

$$\Gamma_L = \left[\begin{array}{c} \left(\Gamma_{L\alpha}^R, \Gamma_{L\alpha}^I \right) \\ \left(\Gamma_{L\beta}^R, \Gamma_{L\beta}^I \right) \end{array} \right] = \left[\begin{array}{c} (\Gamma_0, \Gamma_{R_z}) \\ (\Gamma_{R_y}, \Gamma_{R_x}) \end{array} \right] \otimes \Gamma_\phi \tag{17}$$

where for instance $\Gamma_{L\alpha}^I$ refers to the symmetry of the imaginary part of the $L\alpha$-component. Likewise, the spinor structure of the small components ψ^S in terms of symmetry is given by

$$\Gamma_S = \left[\begin{array}{c} \left(\Gamma_{S\alpha}^R, \Gamma_{S\alpha}^I \right) \\ \left(\Gamma_{S\beta}^R, \Gamma_{S\beta}^I \right) \end{array} \right] = \left[\begin{array}{cc} (\Gamma_{xyz} & , \Gamma_z) \\ (\Gamma_y & , \Gamma_x) \end{array} \right] \otimes \Gamma_\phi = \Gamma_{xyz} \otimes \Gamma_L \tag{18}$$

In the inversion group C_i the irrep Γ_{xyz} is an *ungerade* irrep, and we thus see that the large and small components have opposite parity. Parity alone distinguishes the two possible irreducible corepresentations of the full symmetry group constructed from the antiunitary time reversal operator and the binary groups[12]. This can be seen by consideration of the choice of irrep Γ_ϕ above. Restricting the choice of Γ_ϕ to irreps of the same parity leads only to a redistribution of two-spinor components, whereas a change of parity leads to a qualitatively different spinor.

The 4-component operators span boson irreps, and for an expectation value or a transition moment to be zero the integrand must be totally symmetric. In the matrix representation of a 4-component operator each element comes with the scalar product of two spinors. Even though each spinor spans fermion irreps, their products span boson irreps and consequently the 4-component operator as well. The form of 4-component operators may be deduced (down to a complex phase) from a simple, but powerful observation: In the absence of any external field the Dirac Hamiltonian (13) must be invariant under all possible symmetry operations (unitary or antiunitary) of time and space. This follows from the homogeneity of space and time and from the isotropy of space. The latter implies rotational invariance and the conservation of total angular momentum. The Dirac Hamiltonian commutes with the total angular momentum $\mathbf{j} = \mathbf{l} + \mathbf{s}$. The spin-operator $\mathbf{s}$ is represented by $\frac{1}{2}\rho\boldsymbol{\alpha}$ where $\rho\boldsymbol{\alpha}$ is the 4×4 analogues of the Pauli spin matrices

$$\rho\boldsymbol{\alpha} = (I_2 \otimes \boldsymbol{\sigma}) = \begin{bmatrix} \boldsymbol{\sigma} & 0 \\ 0 & \boldsymbol{\sigma} \end{bmatrix}; \quad \rho = \begin{bmatrix} 0 & I_2 \\ I_2 & 0 \end{bmatrix}; \quad \begin{matrix} [\rho, \beta]_+ &=& 0 \\ [\rho, \alpha] &=& 0 \end{matrix} \quad (19)$$

The total angular momentum operator $\mathbf{j}$ is the generator of infinitesimal rotations. A finite rotation ϕ about an axis represented by the unit vector $\mathbf{n}$ is given by

$$R(\phi, \mathbf{n}) = e^{-i\phi(\mathbf{n}\cdot\mathbf{j})} = e^{-i\phi(\mathbf{n}\cdot\mathbf{l})} e^{-i\phi(\mathbf{n}\cdot\mathbf{s})} = R_r(\phi, \mathbf{n}) R_\sigma(\phi, \mathbf{n}) \qquad \text{as } [\mathbf{l}, \mathbf{s}] = 0 \quad (20)$$

Note from the above relation that the rotation operator splits into one part acting on spatial coordinates and one part acting on spin coordinates. In fact a general point group symmetry operator may be written as

$$\hat{G} = \hat{G}_\mathbf{r}(\phi_\mathbf{r}, \mathbf{n}_\mathbf{r}, p_\mathbf{r}) \hat{G}_\sigma(\phi_\sigma, \mathbf{n}_\sigma, p_\sigma); \qquad p_i = 0, 1 \quad (21)$$

where $\hat{G}_\mathbf{r}(\phi_\mathbf{r}, \mathbf{n}_\mathbf{r}, p_\mathbf{r})$ and $\hat{G}_\sigma(\phi_\sigma, \mathbf{n}_\sigma, p_\sigma)$ act on spatial $\mathbf{r}$ and spin σ coordinates, respectively. They have the form

$$\hat{G}(\phi, \mathbf{n}, p) = \hat{i}^p \hat{R}(\phi, \mathbf{n}) \quad , \quad (p = 0, 1) \quad (22)$$

where $\hat{i}$ represent inversion. One of the major differences between relativistic and non-relativistic formalisms is that in the non-relativistic domain the symmetry operations in Euclidean and spin space are decoupled, and the corresponding two sets of parameters $(\phi, \mathbf{n}, p)$ can therefore be chosen independently, which is not the case in the relativistic domain [14].

The rotation operator for the spin part is straightforwardly established as

$$R_\sigma(\phi, \mathbf{n}) = \cos\frac{1}{2}\phi - i(\rho\boldsymbol{\alpha} \cdot \mathbf{n}) \sin\frac{1}{2}\phi \quad (23)$$

In particular binary rotations about main axes are given by

$$C_2^x = -i\rho\boldsymbol{\alpha}_x; \quad C_2^y = -i\rho\boldsymbol{\alpha}_y; \quad C_2^z = -i\rho\boldsymbol{\alpha}_z \quad (24)$$

We note that $C_2^q C_2^q = -I_4$ for all coordinates q which demonstrates that fermion functions change sign upon a rotation 2π, in contrast to boson functions for which a rotation 2π is equivalent to the identity operation. The fermion phase shift has been verified experimentally in neutron [15, 16, 17] and NMR [18] interferometry.

Representations in spin space of other symmetry operations can be derived using the fact that the operator $(\boldsymbol{\alpha} \cdot \mathbf{p})$ must be invariant under any symmetry operation $\hat{G}$

$$
\begin{aligned}
(\boldsymbol{\alpha} \cdot \mathbf{p}) &= \hat{G} \left(\boldsymbol{\alpha} \cdot \mathbf{p} \right) \hat{G}^{-1} = \sum_i \hat{G}_\sigma \alpha_i G_\sigma^{-1} \hat{G}_\mathbf{r} p_i G_\mathbf{r}^{-1} \\
&= \sum_{ijk} \alpha_j p_k D_{ji}^{\Gamma_{\boldsymbol{\alpha}}}(\hat{G}) D_{ki}^{\Gamma_\mathbf{p}}(\hat{G})
\end{aligned}
\tag{25}
$$

We see that in order for the equality to hold the representation $\Gamma_{\boldsymbol{\alpha}}$ spanned by the Dirac $\boldsymbol{\alpha}$-matrices must be the complex conjugate to that of the momentum operator. The representation matrices of binary groups are all real, so under binary operations the $\boldsymbol{\alpha}$-matrices must transform as the momentum operator $\mathbf{p}$, that is as the Cartesian coordinates. Under space inversion the momentum operator changes sign, so that the spin part of the space inversion operator is determined by

$$
\hat{i}_\sigma \boldsymbol{\alpha} \hat{i}_\sigma^{-1} = -\boldsymbol{\alpha}
\tag{26}
$$

The above relation implies that the inversion operator $\hat{i}_\sigma$ anticommutes with the Dirac $\boldsymbol{\alpha}$-matrices. An obvious choice for the inversion operator is therefore the Dirac β-matrix [see (3)]. We choose

$$
\hat{i}_\sigma = -i\beta
\tag{27}
$$

The imaginary i follows from a consistency argument, as described below. The fact that the Dirac β-matrix appears in the spin part of the inversion operator again demonstrates that the large ψ^L and small ψ^S components have opposite parity. Finally we derive expressions for the operations of reflection in the spin coordinates using the fact that reflections are the product of inversion and binary rotations :

$$
\hat{\sigma}_{yz} = -\beta\rho\alpha_x; \quad \hat{\sigma}_{zx} = -\beta\rho\alpha_y; \quad \hat{\sigma}_{xy} = -\beta\rho\alpha_z
\tag{28}
$$

Two-component analogues of the symmetry operations derived so far are obtained by the substitutions

$$
\rho\boldsymbol{\alpha} \to \boldsymbol{\sigma} \quad \text{and} \quad \beta \to I_2
\tag{29}
$$

In Table 1 we have summarized the two-component representations of the basic eight binary symmetry operations in spin space. It is worthwhile noting that these eight matrices also form the unit vectors of a complex quaternion algebra over a real field[19], as shown in the table. The subset consisting of the identity and the proper rotations are unit vectors for a real quaternion algebra, and this algebra has accordingly found widespread use in the description of rotations in three dimensions[4]. We shall return to this in section 4.

A problematic aspect of the representations derived so far is that they are at

Table 11.1: Mapping between the basic eight binary symmetry operations and the complex quaternion units

Proper rotations			Improper rotations		
$\hat{E}$	$\rightarrow$ I_2	$\rightarrow$ 1	$\hat{\imath}$	$\rightarrow$ $-iI_2$	$\rightarrow$ $-i$
$\hat{C}_2^z$	$\rightarrow$ $-i\sigma_z$	$\rightarrow$ $-\breve{\imath}$	$\hat{\sigma}_{xy}$	$\rightarrow$ $-\sigma_z$	$\rightarrow$ $i\breve{\imath}$
$\hat{C}_2^y$	$\rightarrow$ $-i\sigma_y$	$\rightarrow$ $-\breve{\jmath}$	$\hat{\sigma}_{zx}$	$\rightarrow$ $-\sigma_y$	$\rightarrow$ $i\breve{\jmath}$
$\hat{C}_2^x$	$\rightarrow$ $-i\sigma_x$	$\rightarrow$ $-\breve{k}$	$\hat{\sigma}_{yz}$	$\rightarrow$ $-\sigma_x$	$\rightarrow$ $i\breve{k}$

odds with the conventions of double group theory. The introduction of $\overline{E}$ doubles the order of all rotations. On the other hand, the order of inversion is still taken to be two ($\hat{\imath}^2 = E$), since space inversion commutes with all rotations in ordinary space. In our representation, (27), however, inversion is of order four. Since there is an inherent phase indeterminacy, we could correct this by changing the phase of our inversion operator. However, we can show that operation of inversion *must* be of order four if we want consistency in the representation of spatial symmetry operations.

It is well known that two spins $s = \frac{1}{2}$ couple to a singlet function and the three components of a triplet function. The latter three functions transform as the spherical harmonics Y_{lm} with $l = 1$. By forming the direct product of a 2×2 matrix representation of a symmetry operation (in the two-component case) with itself, we obtain a link to matrix representations of the spherical harmonics with $l = 1$. The direct product gives a 4×4 matrix from which we by a unitary transformation can isolate a 3×3 block representing the corresponding symmetry operation in the basis of spherical harmonics for $l = 1$. The phases for the symmetry operations presented above have been chosen with care so as to obtain agreement with the Condon-Shortley phase convention for spherical harmonics [20]. In the case of inversion, the phase is, however, unambiguous: The direct product of the chosen two-component representation of inversion with itself gives

$$-iI_2 \otimes -iI_2 = -I_4 \tag{30}$$

The identity matrix is invariant under all unitary transformations, and so we obtain $-I_3$ as the representation of inversion in the basis of $\{Y_{1,1}, Y_{1,0}, Y_{1,-1}\}$, as we should (the singlet becomes a pseudoscalar). It is not possible to obtain the same representation starting from a two-component inversion operator of order only two. Altmann [21], in the language of projective representations, sees this discrepancy between representations merely as a choice of gauge (phase). This explanation seems somewhat *ad hoc*. There is a fundamental weakness in the derivation of the behaviour of inversion in double group theory. The extra element $\overline{E}$ is introduced to account for the fact that fermion functions have a behaviour under rotation that is different from rotations in ordinary space. Yet the behaviour of inversion in double group theory is deduced with explicit reference to inversion in ordinary space, which is somewhat inconsistent. It would be interesting to see whether the behaviour of fermion functions under the operations of inversion or reflections can be resolved experimentally.

3.2   Time reversal symmetry

We now turn our attention to antiunitary operators. From (16) we recall that an antiunitary operator $\hat{\mathcal{K}}$ is defined by

$$\left\langle \hat{\mathcal{K}}\phi_1 \mid \hat{\mathcal{K}}\phi_2 \right\rangle = \left\langle \phi_2 \mid \phi_1 \right\rangle = \left\langle \phi_1 \mid \phi_2 \right\rangle^* = \hat{\mathcal{K}} \left\langle \phi_1 \mid \phi_2 \right\rangle \tag{31}$$

The last two terms indicate the antilinearity of antiunitary operators:

$$\hat{\mathcal{K}}\left(a\phi_1 + b\phi_2\right) = a^*\hat{\mathcal{K}}\phi_1 + b^*\hat{\mathcal{K}}\phi_2 \tag{32}$$

It is straightforwardly shown that the product of two antiunitary operators is a unitary operator, which implies that any antiunitary operator can be written as a product of a unitary operator and some antiunitary operator. The simplest choice of an operator to fullfill conditions (31) and (32) is the complex conjugation operator $\hat{\mathcal{K}}_0$. A general antiunitary operator may therefore be written as

$$\hat{\mathcal{K}} = U\hat{\mathcal{K}}_0 \tag{33}$$

where U is a unitary operator. In non-relativistic systems $\hat{\mathcal{K}}_0$ commutes with the Hamiltonian in the absence of external magnetic fields and represents the operation of time reversal [22, 10, 11]. This is straightforwardly seen by letting $\hat{\mathcal{K}}_0$ operate on both sides of the time-dependent Schrödinger equation

$$\hat{\mathcal{K}}_0 \left[i\frac{\partial}{\partial t}\Psi\left(\mathbf{r}, t\right) \right] = \hat{\mathcal{K}}_0 \left[H\Psi\left(\mathbf{r}, t\right) \right]$$

$$\Downarrow$$

$$-i\frac{\partial}{\partial t}\hat{\mathcal{K}}_0\Psi\left(\mathbf{r}, t\right) = i\frac{\partial}{\partial(-t)}\hat{\mathcal{K}}_0\Psi\left(\mathbf{r}, t\right) = H\hat{\mathcal{K}}_0\Psi\left(\mathbf{r}, t\right) \qquad \left[H, \hat{\mathcal{K}}_0\right] = 0 \tag{34}$$

$$\Downarrow$$

$$i\frac{\partial}{\partial t}\left(\hat{\mathcal{K}}_0\Psi\left(\mathbf{r}, -t\right)\right) = H\left(\hat{\mathcal{K}}_0\Psi\left(\mathbf{r}, -t\right)\right)$$

We consider next the form of the time reversal operator $\hat{\mathcal{K}}$ in the 4-component formalism. Momenta are reversed under the operation of time reversal

$$\hat{\mathcal{K}}\mathbf{p}\hat{\mathcal{K}}^{-1} = -\mathbf{p} \tag{35}$$

Using the general form (33) of an antiunitary operator and (25) we therefore find that

$$\hat{\mathcal{K}}\boldsymbol{\alpha}\hat{\mathcal{K}}^{-1} = U\boldsymbol{\alpha}^*U^{-1} = -\boldsymbol{\alpha} \quad \Rightarrow \quad U\left(-\alpha_x, \alpha_y, -\alpha_z\right)U^{-1} = \left(\alpha_x, \alpha_y, \alpha_z\right) \tag{36}$$

Since the Dirac $\boldsymbol{\alpha}$ matrices transform as the coordinates, we identify U as a C_2-rotation about the y-axis (24) and write the time reversal operator as

$$\hat{\mathcal{K}} = -i\rho\alpha_y\hat{\mathcal{K}}_0 = -i\left(I_2 \otimes \sigma_y\right)\hat{\mathcal{K}}_0 \tag{37}$$

Since we are mainly interested in fermion functions, we can alternatively define the time reversal operator by its action on a fermion function ϕ, that is

$$\hat{\mathcal{K}}a\phi = a^*\hat{\mathcal{K}}\phi; \qquad \hat{\mathcal{K}}^2\phi = -\phi \tag{38}$$

We shall use the convention

$$\hat{\mathcal{K}}\phi = \overline{\phi} \tag{39}$$

and denote ϕ and $\overline{\phi}$ as *Kramers partners*. We shall now use the alternative definition (38) to derive the general matrix structure of hermitian operators $\hat{\Omega}_\pm$ that are symmetric(+) or antisymmetric(−) under time reversal [23]

$$\hat{\mathcal{K}}\hat{\Omega}_t\hat{\mathcal{K}}^{-1} = t\hat{\Omega}_t; \qquad t = \pm 1 \tag{40}$$

We consider the matrix representation of $\hat{\Omega}_t$ in a *Kramers restricted basis* which we define as follows: Operate with $\hat{\mathcal{K}}$ on a set of fermion basis functions $\{\phi_p\}$ to generate a complementary basis $\{\overline{\phi}_p\}$. The Kramers restricted basis is then the union of the two sets. The usual relations between matrix elements of $\hat{\Omega}_t$ are easily established

$$\begin{aligned}
\Omega_{\overline{p}\overline{q}} &= \hat{\mathcal{K}}\Omega_{\overline{p}q}^* = \left\langle\hat{\mathcal{K}}\overline{\phi}_q\middle|\hat{\mathcal{K}}\hat{\Omega}\hat{\mathcal{K}}^{-1}\middle|\hat{\mathcal{K}}\overline{\phi}_p\right\rangle &= t\Omega_{pq}^* \\
\Omega_{\overline{p}q} &= \hat{\mathcal{K}}\Omega_{\overline{p}q}^* = \left\langle\hat{\mathcal{K}}\overline{\phi}_p\middle|\hat{\mathcal{K}}\hat{\Omega}\hat{\mathcal{K}}^{-1}\middle|\hat{\mathcal{K}}\phi_q\right\rangle &= -t\Omega_{p\overline{q}}^*
\end{aligned} \tag{41}$$

From these relations follow that the matrix representation of $\hat{\Omega}_\pm$ has the structure

$$\boldsymbol{\Omega}_t = \begin{bmatrix} \mathbf{A} & \mathbf{B} \\ -t\mathbf{B}^* & t\mathbf{A}^* \end{bmatrix}; \qquad \begin{aligned} \mathbf{A}^\dagger &= \mathbf{A}; & A_{pq} &= \Omega_{pq} \\ \mathbf{B}^T &= -t\mathbf{B}; & B_{pq} &= \Omega_{p\overline{q}} \end{aligned} \tag{42}$$

Let us investigate the properties of the above matrix structure. Since $\hat{\Omega}_t$ is hermitian, its matrix may be diagonalized by a unitary transformation, giving real eigenvalues

ε :

$$\begin{bmatrix} \mathbf{A} & \mathbf{B} \\ -t\mathbf{B}^* & t\mathbf{A}^* \end{bmatrix} \begin{bmatrix} \mathbf{c}^\alpha \\ \mathbf{c}^\beta \end{bmatrix} = \varepsilon \begin{bmatrix} \mathbf{c}^\alpha \\ \mathbf{c}^\beta \end{bmatrix} \tag{43}$$

We write out the corresponding matrix equations

$$\begin{aligned} \mathbf{A}\mathbf{c}^\alpha & + & \mathbf{B}\mathbf{c}^\beta & = & \varepsilon \mathbf{c}^\alpha \\ -t\mathbf{B}^*\mathbf{c}^\alpha & + & t\mathbf{A}^*\mathbf{c}^\beta & = & \varepsilon \mathbf{c}^\beta \end{aligned} \tag{44}$$

conjugate both equations and then multiply the first with t and the second with $-t$. This gives

$$\begin{aligned} t\mathbf{A}^*\mathbf{c}^{\alpha*} & + & t\mathbf{B}^*\mathbf{c}^{\beta*} & = & t\varepsilon \mathbf{c}^{\alpha*} \\ \mathbf{B}\mathbf{c}^{\alpha*} & - & \mathbf{A}\mathbf{c}^{\beta*} & = & -t\varepsilon \mathbf{c}^{\beta*} \end{aligned} \tag{45}$$

which can be expressed on matrix form as

$$\begin{bmatrix} \mathbf{A} & \mathbf{B} \\ -t\mathbf{B}^* & t\mathbf{A}^* \end{bmatrix} \begin{bmatrix} -\mathbf{c}^{\beta*} \\ \mathbf{c}^{\alpha*} \end{bmatrix} = t\varepsilon \begin{bmatrix} -\mathbf{c}^{\beta*} \\ \mathbf{c}^{\alpha*} \end{bmatrix} \tag{46}$$

Hence we can conclude the following about the matrix of $\hat{\Omega}_t$ in a Kramers restricted basis

- $t = +1$: $\hat{\Omega}_+$ is *symmetric* with respect to time reversal, and its matrix is doubly degenerate with eigenvectors related by time reversal symmetry

$$\left\{ \begin{bmatrix} \mathbf{c}^\alpha \\ \mathbf{c}^\beta \end{bmatrix}, \begin{bmatrix} -\mathbf{c}^{\beta*} \\ \mathbf{c}^{\alpha*} \end{bmatrix} \right\} \tag{47}$$

To some extent time reversal symmetry recovers the spin symmetry lost in the relativistic domain, but the recovery is only partial. In the non-relativistic domain a totally symmetric (spinfree) operator does not couple two spin orbitals if they have opposite spin. In the relativistic domain we have the weaker relation

$$\left\langle \phi_i \left| \hat{\Omega}_+ \right| \overline{\phi}_j \right\rangle \equiv 0 \quad \text{only if } i = j \tag{48}$$

- $t = -1$: $\hat{\Omega}_-$ is *antisymmetric* with respect to time reversal, and its eigenvectors are also pairwise related by time reversal symmetry (47) with eigenvalues of the same absolute value, but with opposite signs.

Let us now investigate time reversal symmetry in the Dirac equation. It turns out that this is best done by a reordering of the 4-spinors:

$$
\begin{bmatrix} \psi^L \\ \psi^S \end{bmatrix} =
\begin{bmatrix} \psi^{L\alpha} \\ \psi^{L\beta} \\ \psi^{S\alpha} \\ \psi^{S\beta} \end{bmatrix}
\rightarrow
\begin{bmatrix} \psi^{L\alpha} \\ \psi^{S\alpha} \\ \psi^{L\beta} \\ \psi^{S\beta} \end{bmatrix} =
\begin{bmatrix} \psi^{\alpha} \\ \psi^{\beta} \end{bmatrix}
\tag{49}
$$

From the matrix structures derived above we can immediately split the Dirac equation (13) into one part that is symmetric and another part that is antisymmetric under the operation of time reversal

$$
\hat{\mathcal{D}}\psi = \left[\hat{\mathcal{D}}^+ + \hat{\mathcal{D}}^- \right] \psi = 0
\tag{50}
$$

The symmetric part is

$$
\hat{\mathcal{D}}^+ =
\begin{bmatrix}
mc^2 - e\phi & -ic\partial_z & 0 & -ic\partial_- \\
-ic\partial_z & -mc^2 - e\phi & -ic\partial_- & 0 \\
0 & -ic\partial_+ & mc^2 - e\phi & ic\partial_z \\
-ic\partial_+ & 0 & -ic\partial_z & -mc^2 - e\phi
\end{bmatrix}
\tag{51}
$$

where $\partial_\pm = \partial_x \pm i\partial_y$, and the antisymmetric part is

$$
\hat{\mathcal{D}}^- =
\begin{bmatrix}
-i\dfrac{\partial}{\partial t} & ecA_z & 0 & ecA_- \\
ecA_z & -i\dfrac{\partial}{\partial t} & ecA_- & 0 \\
0 & ecA_+ & -i\dfrac{\partial}{\partial t} & -ecA_z \\
ecA_+ & 0 & -ecA_z & -i\dfrac{\partial}{\partial t}
\end{bmatrix}
\tag{52}
$$

We can now explicitly show that the pair of eigenvectors in (47) are related by time reversal symmetry. With reordered spinors (49) the time reversal operator has the form

$$\hat{\mathcal{K}} = -i \left[\sigma_y \otimes I_2 \right] \hat{\mathcal{K}}_0 = \begin{bmatrix} 0 & -I_2 \\ I_2 & 0 \end{bmatrix} \hat{\mathcal{K}}_0 \tag{53}$$

Operating with $\hat{\mathcal{K}}$ on an eigenvector $\mathbf{c}$ we obtain

$$\hat{\mathcal{K}}\mathbf{c} = \hat{\mathcal{K}} \begin{bmatrix} \mathbf{c}^\alpha \\ \mathbf{c}^\beta \end{bmatrix} = \begin{bmatrix} -\mathbf{c}^{\beta*} \\ \mathbf{c}^{\alpha*} \end{bmatrix} = \overline{\mathbf{c}} \tag{54}$$

as required.

Due to the additional structure imposed by time reversal symmetry the number of degrees of freedom of the operator matrix is reduced compared to a general complex hermitian matrix. A $2n \times 2n$ time-symmetric matrix has $n(2n - 1)$ degrees of freedom, whereas a time-antisymmetric matrix of the same dimension has $n(2n + 1)$. The *sum* of these quantities gives $4n^2$, the number of degrees of freedom in the general hermitian matrix. This demonstrates that *any* hermitian matrix[1] may be split into a time-symmetric and a time-antisymmetric part such that the corresponding eigenvalue equation may be written

$$\mathbf{Hc} = [\mathbf{H}_+ + \mathbf{H}_-]\,\mathbf{c} = E\mathbf{c} \tag{55}$$

Operating on this equation with the time reversal operator gives

$$\overline{\mathbf{H}}\overline{\mathbf{c}} = [\mathbf{H}_+ - \mathbf{H}_-]\,\overline{\mathbf{c}} = E\overline{\mathbf{c}} \tag{56}$$

We obtain the curious result that the two matrices $\mathbf{H}$ and $\overline{\mathbf{H}}$ have *identical* eigenvalue spectra, but for each eigenvalue the corresponding eigenvectors $\mathbf{c}$ and $\overline{\mathbf{c}}$ are *strictly* orthogonal. The separation into a time-symmetric and a time-antisymmetric part is not unique.

4   The quaternion Dirac equation

By restricting the Dirac operator to the time symmetric part $\hat{\mathcal{D}}^+$ only, a considerable simplification is possible by the introduction of quaternion algebra. A (real) quaternion number[2] is written as

$$q = \sum_{\Lambda=0}^{3} v_\Lambda e_\Lambda = v_0 + v_1 \check{\imath} + v_2 \check{\jmath} + v_3 \check{k}; \quad v_\Lambda \in \check{\mathbb{R}} \tag{57}$$

---

[1] A hermitian matrix of odd dimension may be extended by a zero row and a zero column

[2] Note that quaternion numbers are not *quaternionic*, just as complex numbers are not *complexionic*. Some uses of quaternion algebra in physics are described in [24, 25].

in which the quaternion units ĭ, ĵ, and ǩ obey the following multiplication rules

$$\check{\imath}^2 = \check{\jmath}^2 = \check{k}^2 = \check{\imath}\check{\jmath}\check{k} = -1 \tag{58}$$

The quaternion units are equivalent in the sense that they may be interchanged by cyclic permutation $\check{\imath} \to \check{\jmath} \to \check{k} \to \check{\imath}$. Thus, in a complex number $a + ib$ the imaginary i may correspond to either ĭ, ĵ, or ǩ without changing its algebraic properties.

When Pauli introduced the spin matrices that bear his name, Jordan pointed out[26] that the properties of imaginary i times the Pauli matrices were identical to that of the quaternion units ĭ, ĵ, and ǩ. Specifically we have the mapping

$$i\sigma_z \leftrightarrow \check{\imath}; \qquad i\sigma_y \leftrightarrow \check{\jmath}; \qquad i\sigma_x \leftrightarrow \check{k} \tag{59}$$

which allows the expression of quaternion numbers by 2×2 complex matrices

$$q = a + b\check{\jmath} \leftrightarrow Q = [s, \mathbf{v}] = s + i\left(\tilde{\boldsymbol{\sigma}} \cdot \mathbf{v}\right) = \begin{bmatrix} a & b \\ -b^* & a^* \end{bmatrix}; \qquad \begin{array}{l} a = s + iv_1; \\ b = v_2 + iv_3; \end{array} \tag{60}$$

with $\tilde{\boldsymbol{\sigma}} = (\sigma_z, \sigma_y, \sigma_x)$, such that

$$q_1 q_2 \leftrightarrow Q_1 Q_2 \tag{61}$$

In the above expression s can be considered the scalar part, $\mathbf{v}$ the vector part. The above matrix representation is analogous to the complex numbers, which may be represented by 2×2 real matrices. Two equivalent representations exist

$$c = a + ib \leftrightarrow \begin{cases} C = \begin{bmatrix} a & b \\ -b & a \end{bmatrix} \\ \\ C' = \begin{bmatrix} a & -b \\ b & a \end{bmatrix} \end{cases} \qquad a, b \in \mathbb{R} \tag{62}$$

The quaternion analogue of C' would be

$$Q' = \begin{bmatrix} a & -b^* \\ b & a^* \end{bmatrix} \tag{63}$$

However we find that $Q'_1 Q'_2 \nleftrightarrow q_1 q_2$. Instead we have $Q'_1 Q'_2 \leftrightarrow q_2 q_1$. This demonstrates a troublesome feature of quaternion numbers, namely that they do not commute under multiplication.

Quaternion algebra allows block diagonalization of the matrix of an operator that is symmetric under time reversal. By comparing (42) and (60) we see that the matrix $\boldsymbol{\Omega}_+$ has a structure identical to that of the 2×2 complex matrix representation

Q of quaternion numbers, which means that it can be expressed in terms of Pauli matrices or quaternion units. Block diagonalization is achieved through the unitary quaternion transformation

$$\mathbf{U}^\dagger \Omega_+ \mathbf{U} = \begin{bmatrix} \mathbf{A} + \mathbf{B}\check{j} & 0 \\ 0 & -\check{k}\left(\mathbf{A} + \mathbf{B}\check{j}\right)\check{k} \end{bmatrix}; \quad \mathbf{U} = \frac{1}{\sqrt{2}} \begin{bmatrix} \mathbf{I} & \check{j}\mathbf{I} \\ \check{j}\mathbf{I} & \mathbf{I} \end{bmatrix} \quad (64)$$

Due to the decoupling of blocks, the transformation leads to an exact reduction of the time reversal symmetric Dirac operator $\hat{\mathcal{D}}^+$ (51) to two-component form, albeit in terms of quaternion algebra (indicated by upper prescript Q):

$$^Q\hat{h}\,^Q\psi = E\,^Q\psi \quad (65)$$

where

$$^Q\hat{h} = \left\{ \begin{bmatrix} mc^2 - e\phi & 0 \\ 0 & -mc^2 - e\phi \end{bmatrix} - c\check{i} \begin{bmatrix} 0 & d_z \\ d_z & 0 \end{bmatrix} - c\check{j} \begin{bmatrix} 0 & d_y \\ d_y & 0 \end{bmatrix} - c\check{k} \begin{bmatrix} 0 & d_x \\ d_x & 0 \end{bmatrix} \right\} \quad (66)$$

The quaternion eigenfunctions $^Q\psi$ are related to the corresponding complex re-ordered 4-spinors (51) by

$$^Q\psi = \begin{bmatrix} \psi^\alpha - \psi^{\beta*}\check{j} \end{bmatrix} \quad \leftrightarrow \quad \begin{bmatrix} \psi^\alpha \\ \psi^\beta \end{bmatrix} \quad (67)$$

The quaternion Dirac operator has an intriguing structure. The scalar potential enters the real part, whereas the kinetic energy part is spanned by the quaternion units $\check{i}$, $\check{j}$, and $\check{k}$. The equivalence of the quaternion units thus parallels the equivalence of the coordinate axes (x,y,z). Note that in the quaternion formulation the time reversal operator is mapped into $-\check{j}$ operating from the right

$$^Q\psi(-\check{j}) = \begin{bmatrix} -\psi^{\beta*} - \psi^\alpha\check{j} \end{bmatrix} \quad \leftrightarrow \quad \hat{\mathcal{K}} \begin{bmatrix} \psi^\alpha \\ \psi^\beta \end{bmatrix} = \begin{bmatrix} -\psi^{\beta*} \\ \psi^{\alpha*} \end{bmatrix} \quad (68)$$

The full Dirac operator (50) may be expressed in terms of *complex quaternions*. Complex quaternions[19, 27] are obtained by replacing the real coefficients in (57) by complex coefficients. The substitutions $\mathbf{A} = i\mathbf{A}'$ and $\mathbf{B} = i\mathbf{B}'$ in (42) establishes the relation $\Omega_- = i\Omega'_+$. In the complex quaternion Dirac operator the time symmetric $\hat{\mathcal{D}}^+$ and antisymmetric $\hat{\mathcal{D}}^-$ parts enter the real and imaginary parts of the complex quaternions, respectively. This arrangement can be understood by observing that a time-antisymmetric matrix obtains a time-symmetric structure with the extraction of imaginary i. In most practical applications, however, we seek stationary states with zero magnetic field, *i.e.* with $\mathbf{A} = 0$ where $\mathbf{A}$ is the vector potential. Nuclear spins and external magnetic fields can usually be handled by perturbation theory.

We shall therefore not pursue this approach here and refer to the literature[28, 29, 30, 31, 32] for details.

5   Discussion and conclusion

The Dirac equation was derived from linearization of the relativistic energy expression (1) using Pauli spin matrices, in particular the relation

$$(\boldsymbol{\sigma} \cdot \mathbf{p})(\boldsymbol{\sigma} \cdot \mathbf{p}) = p^2 \tag{69}$$

which also shows that spin is "hidden" in the non-relativistic Schrödinger equation. Since the quaternion units have the same algebra as the Pauli spin matrices, the quaternion structure is immediately evident. This structure is furthermore reflected in the symmetry properties of the Dirac equation. We have in section 3.1 shown that the eight spatial symmetry operations in spin space of the binary groups map into the unit vectors of a complex quaternion complex algebra over a real field. We have furthermore in section 3.2 shown that quaternion algebra is the natural language of time reversal symmetry.

These links to quaternion algebra have practical implications. Consider the solution of the time-independent Dirac equation for a molecular framework (nuclear spins neglected) in the finite basis approximation. This corresponds to the solution of (50) with $\hat{\mathcal{D}}^- = 0$. The large and small components are expanded into a suitable real basis

$$\psi = \begin{bmatrix} \psi^{L\alpha} \\ \psi^{L\beta} \\ \psi^{S\alpha} \\ \psi^{S\beta} \end{bmatrix} = \begin{bmatrix} \chi^L & 0 & 0 & 0 \\ 0 & \chi^L & 0 & 0 \\ 0 & 0 & \chi^S & 0 \\ 0 & 0 & 0 & \chi^S \end{bmatrix} \begin{bmatrix} \mathbf{c}^{L\alpha} \\ \mathbf{c}^{L\beta} \\ \mathbf{c}^{S\alpha} \\ \mathbf{c}^{S\beta} \end{bmatrix} \tag{70}$$

Note that the two large components are expanded in the same basis $\{\chi_i^L\}$, whereas the two small components are expanded in another basis $\{\chi_i^S\}$. The two basis sets must be related so as to reflect the coupling of the large and small components in the Dirac equation ("kinetic balance" [33, 34]), but this is of no concern for the symmetry aspects considered here.

Solutions are then obtained by diagonalization of the resulting $2N \times 2N$ complex hermitian matrix representation of the Dirac operator where $N = N_L + N_S$ and N_L and N_S are the number of basis functions for the large and the small components, respectively. However, due to the time reversal symmetry of the operator, the matrix can be block diagonalized by a unitary quaternion transformation as in (64) and only one $N \times N$ quaternion hermitian block needs to be considered. This reduces memory and operation count by a factor two.

Further symmetry reductions are possible by invoking spatial symmetry. In the quaternion eigenvalue equation (65) the basis set expansion takes the form

$$
Q_\psi = \begin{bmatrix} \sum_i \chi_i^L(\Gamma_i) \left\{ c_{0i}^L(\Gamma_0) & + & c_{1i}^L(\Gamma_{R_z})\check{\imath} & - & c_{2i}^L(\Gamma_{R_y})\check{\jmath} & + & c_{3i}^L(\Gamma_{R_x})\check{k} \right\} \\ \sum_j \chi_j^S(\Gamma_j) \left\{ c_{0j}^S(\Gamma_{xyz}) & + & c_{1j}^S(\Gamma_z)\check{\imath} & - & c_{2j}^S(\Gamma_y)\check{\jmath} & + & c_{3j}^S(\Gamma_x)\check{k} \right\} \end{bmatrix} \tag{71}
$$

where e.g. $c_{0i}^L(\Gamma_0) = 0$ if $\Gamma_0 \neq \Gamma_i$. Each part of the quaternion coefficients corresponds to a certain position in the Dirac spinor and therefore to a particular boson irrep, as indicated in parenthesis. Consider now a large component basis function of symmetry Γ_0. Clearly the number of non-zero contributions of the corresponding coefficient depends on the number of totally symmetric rotations. We have demonstrated[12] for the binary groups that when there are no totally symmetric rotations in the group, the coefficients of basis functions of *any* boson irrep have only *one* non-zero contribution out of the four real variables in the quaternion coefficient. The coefficients are generally not real since they come with a quaternion unit. However, one may correct for this by *shifting* the quaternion unit over to the basis function

$$
\chi_\Gamma \to e_\Gamma \chi_\Gamma; \quad c_\Gamma^X \to e_\Gamma^* c_\Gamma^X \tag{72}
$$

The binary groups D_2, C_{2v} and D_{2h} have no totally symmetric rotations and by suitable quaternion phase shifts the quaternion eigenvalue problem (65) reduces to a real eigenvalue problem. This means a reduction of memory and operation count by a factor of eight compared to the conventional formulation in (70). For binary groups with only one totally symmetric rotation, that is C_s, C_2 and C_{2h}, suitable quaternion phase shifts give a complex eigenvalue problem. The resulting symmetry scheme provides automatically maximum point group and time reversal symmetry reduction of the computational effort and is straightforwardly extended to the Dirac-Hartree-Fock problem[12]. The symmetry scheme allows one to work with scalar basis functions adapted to boson irreps rather than fermion irreps, just as in non-relativistic calculations. This has the advantage that it allows one to work with conventional integral packages for non-relativistic *ab initio* calculations. A central feature of the scheme is a transfer of symmetry information to the algebra of the problem at hand. The binary groups can be classified into real, complex and quaternion groups, and this classification is evidently linked to the Frobenius theorem referred to in the introduction(for an interesting discussion along these lines see Ref. [35]). The symmetry scheme outlined in this article has been implemented in DIRAC[36, 37], a code for 4-component relativistic molecular calculations, and is to our knowledge the first instance of a symmetry scheme exploiting the full range of this theorem, in the sense that the actual symmetry of the problem decides what algebra to use.

Bibliography

[1] B.N. Kursunoglu and E.P. Wigner, editors. *Reminiscences about a great physicist: Paul Adrien Maurice Dirac.* Cambridge University Press, 1987.

[2] F. G. Frobenius. *Crelle*, 84:59, 1878.

[3] C. S. Peirce. *Amer.Jour.Math.*, 4:225, 1881.

[4] S.L. Altmann. *Rotations, Quaternions, and Double Groups.* Clarendon Press, Oxford, 1986.

[5] G. Aucar, T. Saue, H. J.Åa. Jensen, and L. Visscher. *J.Chem.Phys.*, 110:6208 – 6218, 1999.

[6] B.L. van der Waerden. *Group Theory and Quantum Mechanics.* Springer, Berlin, 1974. (Translation of the German Original Edition: Die Grundlehren der mathematischen Wissenschaften Band 37, Die Gruppentheoretische Methode in der Quantenmechanik. Publisher: Verlag von Julius Springer, Berlin 1932).

[7] H. Weyl. *Z.Phys.*, 56:330–352, 1929.

[8] C.S. Wu, E. Ambler, W. Hayward, D.D. Hoppes, and R.P. Hudson. *Phys.Rev.*, 1413:1413–1414, 1957.

[9] E.P. Wigner. *Group theory and its application to the quantum mechanics of atomic spectra.* Academic Press, New York, 1959.

[10] M. Tinkham. *Group theory and Quantum Mechanics.* McGraw-Hill, New York, 1964.

[11] M. Lax. *Symmetry Principles in Solid State and Molecular Physics.* Wiley and Sons, New York, 1974.

[12] T. Saue and H.J.Aa Jensen. *J.Chem.Phys.*, 111:6211 – 6222, 1999.

[13] H. Bethe. *Ann.Phys.*, 3:133–208, 1929.

[14] J.G. Snijders. *Relativity and pseudopotentials in the HFS method.* PhD thesis, Vrije Universiteit, Amsterdam, 1979.

[15] S.A. Werner. *Phys.Rev.Lett.*, 35:1053–1055, 1975.

[16] H. Rauch, A. Zeilinger, G. Badurek, A. Wilfing, W. Bauspiess, and U. Bonse. *Phys.Lett.*, 54:425–427, 1975.

[17] A.G. Klein and G.I. Opat. *Phys.Rev.Lett.*, 37:238–240, 1976.

[18] M.E. Stoll. *Phys.Rev. A*, 16:1521–1524, 1977.

[19] C.P.Poole and H.A.Farach. *Found.Phys.*, 12:719–738, 1982.

[20] G. Arfken. *Mathematical Methods for Physicists*. Academic Press, San Diego, 1985.

[21] S.L. Altmann and P. Herzig. *Mol.Phys.*, 45:585–604, 1982.

[22] E.Wigner. *Nachrichten der Akad. der Wissensch. zu Göttingen,II*, pages 546–559, 1932.

[23] H.J.Aa Jensen, K.G. Dyall, T. Saue, and K. Fægri. *J.Chem.Phys.*, 104:4083–4097, 1996.

[24] Max Jammer. *The Conceptual Development of Quantum Mechanics*. McGraw-Hill, New York, 1966.

[25] G.M. Dixon. *Division Algebras*. Kluwer Academic, 1994.

[26] Footnote 2 on page 607 in [38].

[27] L.E. Dickson. *Linear Algebras*. Cambridge University Press, London, 1914.

[28] C.Lanczos. *Z.Phys*, 57:474–483, 1929.

[29] C.Lanczos. *Z.Phys*, 57:484–493, 1929.

[30] A.W.Conway. *Proc.Roy.Soc. (London) A*, 162:145–154, 1937.

[31] P.Rastall. *Rev.Mod.Phys.*, 36:820–832, 1964.

[32] K.Morita. *Prog.Theor.Phys.*, 70:1648–1665, 1983.

[33] A.D. McLean and Y.S. Lee. In R. Carbo, editor, *Current Aspects of Quantum Chemistry 1981*, volume 21, pages 219–238, Amsterdam, 1982. Elsevier.

[34] R.E. Stanton and S. Havriliak. *J.Chem.Phys.*, 81:1910–1918, 1984.

[35] F.J. Dyson. *J.Math.Phys.*, 3:1199–1215, 1962.

[36] Dirac, a relativistic ab initio electronic structure program, Release 3.1 (1998), written by T. Saue, T. Enevoldsen, T. Helgaker, H. J. Aa. Jensen, J. Laerdahl, K. Ruud, J. Thyssen, and L. Visscher. See http://dirac.chem.sdu.dk.

[37] T.Saue, K.Fægri, T.Helgaker, and O.Gropen. *Mol.Phys.*, 91, 1997.

[38] W. Pauli. *Z.Phys.*, 43:601–623, 1927.

Lecture Notes in Chemistry

For information about Vols. 1–34
please contact your bookseller or Springer-Verlag

Editorial Policy

This series aims to report new developments in chemical research and teaching - quickly, informally and at a high level. The type of material considered for publication includes:

1. Preliminary drafts of original papers and monographs

2. Lectures on a new field, or presenting a new angle on a classical field

3. Seminar work-outs

4. Reports of meetings, provided they are
   a) of exceptional interest and
   b) devoted to a single topic.

Texts which are out of print but still in demand may also be considered if they fall within these categories.

The timeliness of a manuscript is more important than its form, which may be unfinished or tentative. Thus, in some instances, proofs may be merely outlined and results presented which have been or will later be published elsewhere. If possible, a subject index should be included. Publication of Lecture Notes is intended as a service to the international chemical community, in that a commercial publisher, Springer-Verlag, can offer a wider distribution to documents which would otherwise have a restricted readership. Once published and copyrighted, they can be documented in the scientific literature.

Manuscripts

Manuscripts should comprise not less than 100 and preferably not more than 500 pages. They are reproduced by a photographic process and therefore must be submitted in camera-ready form according to Springer-Verlag's specifications: technical instructions will be sent on request.

The text area should take care of the page length and width (12.2 x 19.3 cm when you use a 10 point font size, 15.3 x 24.2 cm for a 12 point font size).

Authors receive 50 free copies and are free to use the material in other publications.

Manuscripts should be sent to one of the editors or directly to Springer-Verlag, Heidelberg.